TIME EXPANSION

EXPERIENCES

ALSO BY STEVE TAYLOR

THE PSYCHOLOGY OF TIME PERCEPTION
AND THE ILLUSION OF LINEAR *TIME*

TIME

EXPANSION
EXPERIENCES

Steve Taylor PhD

WATKINS
1893

This edition first published in the UK and USA in 2024 by
Watkins, an imprint of Watkins Media Limited
Unit 11, Shepperton House
89-93 Shepperton Road
London
N1 3DF

enquiries@watkinspublishing.com

1 2 3 4 5 6 7 8 9 10

Typeset by JCS Publishing Services Ltd

Printed and bound by CPI Group (UK) Ltd,
Croydon, CR0 4YY

A CIP record for this book is available from the British Library

ISBN: 978-1-78678-846-7 (Paperback)
ISBN: 978-1-78678-847-4(eBook)

www.watkinspublishing.com

CONTENTS

INTRODUCTION
WHEN SECONDS TURN INTO MINUTES

Ten years ago, I had a car crash. I was driving in the middle lane of a busy motorway. My wife was in the passenger seat. A truck pulled out from the inside lane and hit the side of our car, spinning us around, and then hitting us again.

Everything went into slow motion. I heard the sound of the first impact and asked my wife, "What was that noise?" After a long pause, the car started to spin. I looked behind, and the other cars seemed to be moving extremely slowly, almost as if they were frozen. Time seemed suspended, as if a "pause" button had been pressed. I felt as though I had a lot of time to observe the whole scene and try to regain control of the car. I was surprised by how clear and vivid everything became and how much detail I was able to perceive. I could see the long rows of cars stretching back through the lanes behind us and the shocked faces of the drivers right behind.

I was also surprised at my calmness. Rather than panicking, I thought clearly and methodically about the situation. I tried to regain control of the car, gripping the steering wheel, pressing down on the brake and trying to turn toward the hard shoulder (the empty inside lane of the motorway, reserved for breakdowns). But the car kept on spinning for what seemed at least half a minute, although in reality it was probably no more than three or four seconds. We spun toward the hard shoulder then hit a crash barrier and finally stopped.

At that point, time seemed to return to its normal speed. I looked at my wife and then at my own body and was amazed that neither of us had serious injuries. I looked around the interior of the car, which didn't show any signs of damage (although the exterior was so badly mangled that it had to be scrapped).

I felt an incredible sense of elation and relief. If the car had moved in any other direction, straight ahead or toward the right, there would undoubtedly have been a major accident. We would probably have been killed, together with other drivers. I'm not sure whether my actions helped the car to move toward the hard shoulder or if the movement was purely random. However, it certainly felt as though my calmness and my methodical thoughts and actions helped us survive a potentially fatal incident.

This is an example of what I call a "time expansion experience" (TEE, for short). Time expansion experiences are so common in accidents and emergencies that there is a good chance that you, the reader, have had one – most likely in a car accident, but perhaps also in a fall, a medical emergency or a sudden shocking event such as a robbery, a violent attack or an earthquake.

However, TEEs don't just occur in emergency situations. In my research, I have found that they can occur in traumatic or tragic events such as bereavement, diagnosis of cancer and the breakdown of a marriage. They may occur in the context of sports and games, when a person enters a deep state of absorption, sometimes referred to as the "Zone". TEEs may also occur in unusual states of consciousness induced by deep meditation or by psychedelic substances such as LSD or ayahuasca.

Paradoxically – as they often involve dangerous, even life-threatening situations – TEEs are usually extremely positive experiences. In my research, 83 per cent of people report feeling calm, even serene. Despite danger and trauma, they feel free from pain and anxiety. There is also often a sense of alertness and clarity. People feel intensely present, with a heightened awareness of their surroundings. The world looks more real and beautiful than normal. In emergency situations, they usually feel – as I did – able to think clearly, with the opportunity to deliberate over their predicament and take preventative action.

In accident-related TEEs, time typically appears to expand by many orders of magnitude – usually from 10 to 40 times, so that a period of a few seconds may seem to stretch for minutes. TEEs in other contexts may be even more dramatic and extraordinary. A person in a deep hypnotic trance may – if given appropriate instructions by the hypnotist – experience a few seconds as hours, demonstrating this by performing complex and detailed intellectual feats that would normally only be possible over a long period of time. A similar type of time expansion may occur under the influence of LSD or DMT.

Even more extreme, in near-death experiences – when a person skirts dangerously close to death or actually does clinically die for a brief period, before resuscitation – time slows down so radically that a long series of complex experiences (equivalent to several hours in normal time) may unfold in a few seconds of normal time. Sometimes this includes the strange phenomenon of a "life review" in which a person views every single experience of their life (and sometimes apparently their future too) in a panorama playing out before them, like a film. The life review is perhaps the most radical time expansion of all. How is it possible to relive a whole lifetime of experiences in a matter of seconds?

This is just one of the questions I will attempt to answer in this book. Other central questions include: what are the main situations in which TEEs occur? Why do they occur in certain situations but not in others? Why are some people more prone to them than others? How can we explain TEEs? What do they tell us about consciousness, the human mind and the nature of reality itself? Is it possible for us to consciously induce TEEs or, more generally, to slow down our everyday experience of time?

My Own Perspective

I have been fascinated by time perception for most of my adult life. Looking back to my childhood, I can remember situations when time seemed much more expansive, such as short car journeys that seemed to last for hours. I have a clear memory of finishing primary school aged 11, knowing that I would be starting secondary school in six weeks, after the summer holidays. I started to ponder over secondary school, wondering what it would be like and if I should worry about it. But then I told myself, "Well, there's no point thinking about it, as it's so far ahead in the future." The six-week period that stretched ahead of me seemed so expansive that it was probably the equivalent of six months of my adult life.

As an adult, I noticed that time seemed to pass slowly when I was bored and to speed up when I was absorbed in an activity like reading, playing music or socializing with friends. Perhaps most notably, I noticed that time seemed to stretch when I travelled to unfamiliar places. In my early twenties, I was a musician and toured Germany with my band. I met a girl and decided to leave the band to be with her. I moved to eastern Germany, just a year after the fall of the Berlin Wall, and everything seemed exhilaratingly different and strange. My life changed radically. Besides the hyper-reality of a new environment, I was living with a partner for the first time. I joined a new band and started to give English lessons to augment my income. Almost no one spoke English (in East Germany people learned Russian at school, as English was the language of the capitalist West) so I had to work hard to learn German, in order to communicate.

After eight months, I came back to the UK on holiday and felt like I had been away for more like eight years. I felt like a Roman solider returning home after years in a far-flung corner of the empire. I was genuinely shocked that the same people were working in the same shops, and that my friends were doing the same jobs. I felt that I had been away for so long that major changes should have occurred in people's lives. Similarly, a couple of years later (by which time eastern Germany had become familiar to me) I travelled around India for six weeks, where I had so many intense experiences that my nervous system could barely cope. On my return, I felt as if I had been away for months.

Several years later – after I had abandoned music and made a new career in the academic world – I decided to investigate time perception as a psychologist. This led to my earlier book *Making Time*, published in 2007, where I attempted to explain why time seems to pass at different speeds in different situations. (I'll summarize some of my findings from that book in Chapter 1.) Afterwards, I put the topic of time to one side, and spent the next few years investigating other areas of psychology. However, the car crash described above refocused my attention on time perception. I began to collect reports of other people's TEEs, and to systematically analyse them, examining their contexts and characteristics. In 2020, my first study of TEEs – entitled "When Seconds Turn to Minutes: Time Expansion Experiences in Altered States of Consciousness" – was published in the *Journal of Humanistic Psychology*. This article was a study of 96 reports of TEEs. I also published several popular articles on the topic, which led to people sending me many more reports.

An Overview of This Book

In this book, I share the findings of my research. The first chapter is a general introduction to the topic of time perception (feel free to skip this chapter if you're already familiar with the topic, or would simply like to dive straight into the world of time expansion experiences). In this chapter, I deal with time-honoured (pun intended) questions such as: why does time seems to speed up as we get older? Why does it seem to speed up when we're enjoying ourselves or in a state of absorption? Why does it seem to slow down in unfamiliar environments? We look at some of the theories that psychologists have developed to explain these different experiences. Then I present my concept of four fundamental "laws" of psychological time and two principles of psychological "relativity" that underpin them.

Following this, from Chapter 2 onward, we look specifically at TEEs. We explore the different contexts in which they occur and I describe some of the most interesting and dramatic cases from my research. We begin with accidents and emergencies (including traumatic life events). I argue that such TEEs are real experiences that happen in the moment. They are not – as has been suggested by some researchers – an illusion of recollection, caused by an increased number of memories created in unusual situations.

In Chapter 3, we move on to TEEs in sports and games. I suggest that the ability to slow down time is the key to sporting greatness, and that the highest-level athletes may operate in a different "timeworld" to their competitors. In Chapter 4, we examine TEEs in the context of psychedelic and spiritual experiences. Beginning with this chapter – and continuing through later ones – we also discuss what I call "time cessation experiences" (or TCEs). In one type of time cessation experience, time simply disappears. We step out of time in the same way that a swimmer might step out of a river. In another variant, time becomes *spatial* rather than linear. The future and past seem to exist in parallel with the present. In these moments, it is possible to glimpse past and future experiences because they are simply *there* in front of us, as a part of a wide-ranging landscape.

In Chapter 5, we discuss TEEs and TCEs in near-death experiences, moving on (in Chapter 6) to the life review. In Chapter 7, we discuss some potential explanations of TEEs and TCEs. Is it possible to explain them in terms of unusual neurological activity or as an adaptive ability that evolved because it helped early humans to survive in dangerous situations?

A clue to understanding the experiences is that they all occur in altered states of consciousness, when our normal psychological functioning significantly changes. It is well known that psychedelics and deep meditation induce altered states of consciousness – they change our perception of our surroundings, our sense of identity, our ability to concentrate and process information, and so on. Athletes sometimes enter the altered state of the "Zone" while competing, due to intense concentration. And I argue that accidents and emergencies may have the same effect. The sudden shock of an accident may also jolt us into an altered state of consciousness.

Altered states of consciousness always bring changes to time perception. (This includes hypnotism and dementia, which we discuss in detail in Chapter 7.) There are some mild altered states of consciousness in which time seems to speed up, such as the "flow" state, when we are intensely absorbed in an activity. There are also some drug-induced altered states – usually with sedatives or narcotics that reduce our awareness and mental activity – that have a time-contracting effect. But in most altered states, especially more intense ones, time slows down significantly. There seems to be a correlation between the intensity of an altered state and the intensity of a TEE – the more our state of consciousness alters, the slower time seems to pass. As we will see, there are even some radically altered states in which time seems to disappear completely. This is when the TCE occurs.

There is one particular feature of altered states that generates TEEs and TCEs: the weakening of our normal sense of separation, the boundary that seems to exist between us and the world "out there". The softer this boundary grows, the more time slows down. And when the boundary disappears, time disappears too. All of this leads to the startling idea that our normal sense of linear time may simply be a construct, produced by our normal state of consciousness – and in particular, by our sense of separation.

In Chapter 8, we look at evidence from studies of precognition (the ability to sense and anticipate future events) that supports the notion of "spatial" time. In Chapter 9, we explore several concepts and theories from modern physics (including the concept of "eternalism") that also support this notion, implying that our normal sense of linear time is illusory.

I conclude by discussing what all of this means to us as individuals, in our day-to-day lives. As we know that time is so elastic, is it possible for us to consciously manipulate it, to slow it down at will or even speed it up in certain situations? Can we consciously produce TEEs or even TCEs? Is it possible to live our lives on the basis that linear time is an illusion?

I hope that time will not pass slowly for you as you read this book. And I'm confident that it won't. Time is such a fascinating topic because it is the basis of our lives. We live in time, just as fish swim in water. So, to examine time is to examine life itself. And to understand time is to understand ourselves.

CHAPTER 1
WHY TIME SEEMS TO PASS AT DIFFERENT SPEEDS

An Overview

It's doubtful that modern psychology would exist – at least in its present form – without William James, who is often referred to as the "founding father" of American psychology. As well as offering the first ever psychology courses in the US at Harvard University, he wrote *The Principles of Psychology*, the first detailed textbook of psychology, published in two volumes in 1890.

James was born in 1842, 14 years before Freud, but his approach to psychology seems more contemporary than the latter's. Freud's approach was influential for many decades, yet now seems outmoded and even somewhat bizarre, with its emphasis on syndromes such as penis envy and castration complex. In contrast, James's psychology seems strikingly modern. In *The Principles,* James discusses most of the topics that preoccupy modern psychologists, such as memory, attention, perception, consciousness and the relationship of the mind to the brain.

One of the central chapters of *The Principles* is on time perception. James was a well-travelled man, especially for his era. Apparently viewing a boat journey of several days as a minor inconvenience – equivalent to how we might view a few hours on a plane nowadays – he made many extended visits to Europe. These travels no doubt informed his insights about time perception. As James reflected, "rapid and interesting travel" results in the same "multitudinous, and long-drawn-out" time perception as childhood.[1] He speculated that this was due to intensity of perception. In other words, the more intense perception becomes, the more "long-drawn-out" time becomes. James suggested that this was also why time seems to speed up as we get older. As he wrote, "in youth, we have an absolutely new experience, subjective or objective, every hour of the day . . . but as each passing year converts some of this experience into automatic routine which we hardly note at all, the days smooth themselves out in recollection to contentless units, and the years grow hollow and collapse."[2]

Another of James's insights was that time as we experience it in the moment is sometimes different to how we perceive it *in retrospect*. As he put it, "In general, a time filled with varied and interesting experiences seems short in passing, but long as we look back. On the other hand, a tract of time empty of experiences seems long in passing, but in retrospect short."[3] James felt that this was an effect of memory. Interesting experiences create lots of memories, while empty periods of time create few. He offers the example of a week of travel, which generates so many memories that in retrospect it seems like three weeks, and the counterexample of a month of sickness, which "hardly yields more memories than a day".[4]

Time and Information

Over the last few decades, James's insights have been confirmed by researchers and developed further by theorists. In fact, research over the last few decades has provided a fairly clear picture of the different factors that affect time perception.

Perhaps the most well-established finding is a link between time perception and information processing. The more information our minds process, the slower time seems to pass. This fits with James's insight that new experiences slow down time perception. When we travel abroad – or when we are children – our minds process more information, simply because so much around us is unfamiliar. We pay much more attention to our surroundings. The buildings and scenery seem more vivid and beautiful. We pay attention to the taste of the unfamiliar food, the sound of the strange accents or foreign languages, the new smells and so on – all of which stretches our experience of time. But when we go back home – or when we become adults – we pay less attention to our surroundings. Our minds seem to switch off to the familiar sights, smells and sounds. We live in our normal environments, according to normal routines. As a result, we process less information and time seems to speed up.

The link between information processing and time perception was first established empirically by the American psychologist Robert Ornstein. In a series of experiments beginning in the late 1960s, participants were played tapes containing various kinds of auditory information, such as clicking noises and household sounds. As the amount of information increased, participants estimated periods of time as longer. Ornstein found that this applied to complexity of information too. When they were asked to examine different drawings and paintings, the participants with the most complex images estimated the longest period of time.[5] Further experiments along similar lines by other psychologists have found that people overestimate time periods that contain unfamiliar or an increased variety of information and when information is segmented (rather than presented in a steady flow).[6]

One simple way to demonstrate the link between information and time is to listen to contrasting pieces of music. Often when I do university lectures or public talks on time perception, I play two pieces of music, asking my audience to estimate their duration. I usually play an ambient piece by

Brian Eno, which glides by gracefully with few notes and little variation. This is followed by an excerpt from a frenetic piano concerto by Rachmaninov, full of rapid cascading notes. I play the pieces for the same amount of time – usually 3–4 minutes, depending on how much time is available. Without fail, the audience estimates more time for the Rachmaninov piece, usually between a third and a half longer. (Interestingly, I have found that if people listen to ambient uneventful music for longer periods of time – say, ten minutes – the time contraction becomes less significant, probably because the time-contracting effect of lack of information is offset by the time-expanding effect of boredom.)[7]

More recently, psychologists have found that our experience of time doesn't just depend on *what* we perceive, but also on *how* we perceive. The most important factor here is "perceptual clarity".[8] Imagine you wake up after a long lie-in on a Sunday morning, feeling refreshed. You make a cup of coffee and wander around your garden in the sunshine. You're struck by the brightness and beauty of the plants and flowers. You can hear birds singing and children playing in a nearby park. You gaze above and are struck by the endless space and stillness of the sky . . .

This is typically a "time-expansive" situation, because of the clarity of your perception. Clear perception opens our awareness to our surroundings, allowing us to absorb more impressions and sensations, in the same way that a clean window enables us to see more of the scene outside. When perception is clear, a veil of familiarity seems to fall away from our surroundings and we perceive phenomena as if they are fresh and new.

One reliable way of cultivating clear perception is meditation. A good meditation stills the mind, slowing down our restless thoughts. With our minds quieter, our perceptions become richer and our surroundings seem more vivid. After a good meditation, I often find myself noticing things that I don't normally pay attention to. I look around the room and notice images and patterns on T-shirts or mugs or carpets, patterns of rain on the window or the swaying motion of trees outside. In such moments, time seems to become more expansive too.

Time and Mood

Sometimes members of the public accuse academic researchers of wasting their time trying to establish facts that are already obvious to common sense. A couple of years ago, I did an interview about time perception for a popular German newspaper. The online version attracted a few disparaging comments such as "Do we really need a psychologist to tell us this?" and "Whose money has he been wasting on this research?" (Thanks a lot, guys! Fortunately, there were some positive comments too, which offset the damage to my self-esteem.)

It's true that there are many common pieces of folklore telling us "time flies when you're having fun" or "a watched pot never boils." These express such fundamental human experiences that there are equivalent sayings in many other languages, such as the French: *"le temps passe vite quand on s'amuse"* or the Dutch *"tijd vliegt als je het leuk hebt."* Nevertheless, in my view it's important for researchers to investigate common assumptions and ascertain if they are based in fact.

So what does research tell us about the notion that *tijd vliegt als je het leuk hebt* (time flies when you're having fun)? It has certainly been established that time perception is linked to mood. Put simply, when we're in a good mood, time seems to pass quickly; when we're in a bad mood, it seems to pass slowly. States such as happiness, excitement and absorption make time pass quickly, while states such as boredom, depression or anxiety make time pass slowly. It's a shame this isn't the other way round, you might think.

It's almost as if an evil deity is playing perverse tricks on us, extending our suffering and shortening our pleasure.

A study by the Israeli psychologists Dinah Avni-Babad and Ilana Ritov showed that air travellers who fly regularly experienced "a swifter passage of time" than those who only fly occasionally, presumably because the latter felt less secure and comfortable.[9] Another study by psychologists at Penn State University found that people who had just given up smoking significantly over-estimated time periods, presumably due to their discomfort and inability to concentrate.[10] Similarly, in a study conducted by Marc Wittmann and Martin Paulus, patients in treatment at a drug clinic (with a history of dependency on cocaine and methamphetamine) estimated a 53-second interval as lasting 24 seconds longer than other people, and also reported weeks and months as lasting longer.[11] Other laboratory tests have shown that when people are shown threatening or emotionally charged images, they tend to overestimate time, compared to neutral images. For example, erotic scenes or pictures of accidents seem to stretch time, compared to images of grazing cows.[12]

One of the world's leading contemporary researchers on time perception is Dr Ruth Ogden, who is carrying out exciting research with her team at Liverpool John Moores University. In a study that investigated people's experiences of time during the UK's Covid lockdown, Ogden found a link between the passage of time and levels of social satisfaction and stress. People with a higher level of social satisfaction and less stress reported a swifter passage of time during lockdown, presumably because of their more positive moods.[13] In another study led by Ogden, 398 chronic pain sufferers were asked how they perceived time when in pain, compared to when pain abated. Eighty-six per cent reported that time passed more slowly when their pain was at its worst (40 per cent reported that time passed "extremely slowly"). As one person stated, "It just drags, as if I'm swimming through treacle . . . Even waiting for a few seconds for the microwave to ping or the kettle to boil seems to last forever at times."[14]

One important factor here is simply *awareness* of time. Time goes quickly when we forget about time and drags when we are conscious of it. Time flies for comfortable airplane passengers partly because they forget about the duration of their journey. They can just read, watch films and chat to friends, oblivious to the passing of time. In contrast, anxious passengers continually think about the duration of the flight, checking how long is left and imagining landing safely, unable to immerse their attention in a book or film. The nearer the plane draws to its destination, the more time-conscious they become, and the slower time seems to pass. The same applies to pain sufferers. They become more conscious of time because they are unable to focus their attention on anything else.

This implies that *impatience* stretches time, which makes sense when we consider that impatience essentially *is* awareness of time. Impatience means being acutely aware of a desired future event and of the gulf of time that separates us from it. Desperate to reach the future, we become dissatisfied with the present. And of course, this has the counterproductive effect of expanding the gulf of time. This is the essence of the saying: "a watched pots never boils." (Or as they say in Italian, "*Pentola guardata non bolle mai.*")

What Determines Our Time Perception?

What about the psychological and neurological processes that determine our experience of time? Unfortunately, there still isn't a great deal of clarity on this. One popular theory is the "scalar expectancy theory", originally developed in 1977 by psychologist John Gibbon. This suggests that our sense of time passing stems from a kind of pacemaker (or internal clock) in the brain, which produces regular pulses. These are counted by an "accumulator" and stored in our memory. At the end of a period of time, we remember how many pulses the pacemaker has made and so have a rough idea of have much time has passed.[15]

According to this theory, time seems to slow down when the pacemaker pulses more frequently, and to speed up with fewer pulses. Gibbon thought that his theory could account for different experiences of time in animals, whose pacemakers may pulse faster or slower than ours. Animals that seem to live for a short time from our perspective – such as insects – may actually experience a sizeable life span, if their pacemakers pulse very quickly. An insect that lives for just a few weeks may experience a lifespan equivalent to years of human time.

Unfortunately, Gibbon's theory doesn't explain why human beings' time perception is so variable. Why would new experiences and negative states of mind increase the pacemakers' pulse rate, bringing a slower passage of time? Why would familiarity and pleasant states of mind decrease the rate, causing a faster passage? To try to address this issue, in the 1990s, psychologists Dan Zakay and Richard Block added another factor to Gibbon's original theory: attention. They suggested that the mind has an "attentional gate" that can be wide or narrow, allowing fewer or more signals to pass through. When we are aware of time passing, the gate is wider and more pulses pass through, whereas forgetting about time closes the gate and reduces the pulses.[16] Perhaps the "attentional gate" concept could be applied to new experience and familiarity too. It may be that when we become more aware of our surroundings, more perceptions are allowed through the gate.

However, as yet – even after 45 years – no one has yet discovered if the "pacemaker-accumulator" actually exists, or which brain processes it might relate to. In fact, researchers have not located any type of internal neurological clock that underlies our perception of time.

After many years researching the neurological basis of time perception, the neuroscientist Dean Buonomano has cited a "lack of empirical support for the internal clock model",[17] with no "converging evidence that there is any master circuit within the brain responsible for telling time on the scale of hundreds of milliseconds to a few seconds".[18] Buonomano and his team have identified certain brain processes that seem to be associated with the discrimination of intervals or the integration of information across time. However, these processes aren't very precise, and just seem to record time, rather than measure it. As Buonomano puts it, they are more like "the digital display of a wristwatch" rather than the "quartz crystal" inside it.[19]

Another possibility is that the basis of time perception is not in the brain but the body. Another of the world's leading time perception researchers, Marc Wittmann, has suggested that we measure time through our awareness of physical processes. Or as he puts it, "body signals provide the basis for subjective time."[20] This is a slight variant of the information processing theory, in which the main source of information is our bodies and emotions. The more signals we receive from our bodies – that is, the more aware we are of physical processes and emotions, such as when we are in pain or in a waiting situation – the slower time seems to pass. To put it more simply, the more aware we are of our bodies, the more aware we are of time and the slower it seems to pass. But when we become less aware of physical processes and emotions – such as in a state of absorption – we become less aware of time and it seems to pass quickly. Building on the work of the American neuroscientist Bud Craig, Wittmann highlights a link with a part of the brain called the insular cortex, which appears to be strongly associated with both time perception and the processing of signals from the body. This link is supported by research indicating that when the insular cortex is damaged after a stroke, patients' judgement of duration became less accurate.[21]

Wittmann's approach sounds promising. However, this theory would still need to describe *how* the insular cortex generates our sense of duration. More problematically, the theory struggles to account for time expansion experiences. In TEEs, the correlation between bodily processes and time perception seems to break down. People don't appear to be especially aware of physical processes. The number of "body signals" that we experience doesn't seem to increase massively. People almost always feel calm, even relaxed, as if their metabolism has slowed down rather than speeded up. According to the above theory, this would result in a speeding up of time, rather than a radical slowing.

Significantly, this overall lack of clarity about time perception mirrors our neurological uncertainty about consciousness itself. When the first brain-scanning technologies were developed, many scientists believed they would soon find out how the brain generates our subjective experience. However, their progress was disappointing and, decades later, the "neurological correlates" of consciousness remain elusive. In 1998, the neuroscientist Christof Koch bet the philosopher David Chalmers that in 25 years scientists would have a clear idea of the "neural signature of consciousness". In 2023, Koch admitted that he had lost the bet and handed Chalmers a case of wine. Many scientists now believe that consciousness doesn't emerge from any particular neurological structure or process but is somehow associated with the brain as a whole. Other scientists have responded by adopting alternative approaches such as panpsychism (which suggests that consciousness is a fundamental feature of the universe) or by claiming that consciousness is an illusion.[22]

Since time is closely associated with consciousness – and as we will see, since altered states of consciousness bring different time perception – our uncertainty about time perception isn't surprising. If we can't explain the neural basis of consciousness, why would we expect to understand time in neurological terms?

Four Laws of Psychological Time

In my view, the best way to describe our general experience of time is to think in terms of "laws" of psychological time. This is the approach I took in *Making Time* (although I have since modified the laws that I developed in that book). In the rest of this chapter, I'll describe four essential laws and explain them in terms of the factors discussed above (such as information processing and mood). The fourth law is distinct from the first three and relates to different factors, so it will be discussed in detail in later chapters.

1. Time seems to speed up as we get older.

When I enquire at talks and lectures, it's rare that anyone disagrees with the above statement. Even teenage students generally feel that time is moving faster than when they were children, while twenty-somethings agree that time is moving faster than when they were teenagers. In a recent study of 918 adults led by Ruth Ogden, 77 per cent of respondents agreed that Christmas seems to arrive more rapidly each year. (14 per cent were neutral on the issue, while only 9 per cent disagreed.) Interestingly, Ogden's co-researchers asked an Iraqi sample the same question about Ramadan and received a very similar response.[23]

At lectures and talks, I often ask if anyone has any theories about why time seems to speed up as we get older. Almost always, a member of the audience raises their hand and says something along the lines of, "My idea is that it's proportional to the whole of your life. When you're a child, each year is a massive proportion of your life, so it seems like a long time. But when you're old, each year is a small proportion of your life, so it doesn't seem as significant." I'm reluctant to tell people that their idea isn't original but feel obliged to point out that this theory was first put forward almost 150 years ago by the French philosopher Paul Janet.

As William James paraphrases Janet in *The Principles of Psychology,* "the apparent length of an interval at a given epoch of a man's life is proportional to the total length of the life itself. A child of 10 feels a year as 1/10 of his whole life – a man of 50 as 1/50, the whole life meanwhile apparently preserving a constant length."[24]

There is some sense to this theory – it would explain why the speed of time seems to increase so gradually and evenly, with almost mathematical consistency. But I don't think it can *fully* account for the speeding up of time that we experience. The "proportional theory" assumes that we always experience our lives *in toto*, perceiving each day, week, month or year in relation to the whole of the time we've been on the surface of this planet. But we don't live our lives like this. We live in terms of much smaller periods of time, from hour to hour and day to day, dealing with each period on its own merits. We may relate our present time to the recent past, but not usually to our whole lives.

I agree with William James – who was also doubtful of the proportional theory – that the main reason why time passes so slowly for children is because "we have an absolutely new experience, subjective or objective, every hour of the day" and that time speeds up with increasing age because "each passing year converts some of this experience into automatic routine, which we hardly note at all."[25] In other words, the increased speed of time is linked to information processing. Because they have so many new experiences, children process a massive amount of perceptual information.

However, I don't think it's just a question of new experiences, but also of children's unfiltered and intense perception of the world. This makes their surroundings appear more real, brighter and fresher, full of intricate details that adults don't notice – tiny cracks in windows, insects crawling across the floor, patterns of sunlight on the carpet and so on. For children, all this information stretches time. However, as we get older, we have progressively fewer new experiences. Equally importantly, our perception of the world become more automatic. We become progressively de-sensitized to our surroundings. As a result, we absorb gradually less information, which means that time passes more quickly. Time is less stretched with information. (There is a further, more subtle reason why time passes so slowly for children, which we will examine in more detail later: because they don't have a strong sense of self or a sense of separation from the world. As the sense of self – and of separation from the world – develops, the passage of time speeds up.)

This process isn't completely inevitable. There are certain things we can do to resist the process of time speeding up. The most obvious is to keep introducing newness into our lives by travelling to new places, learning new hobbies, meeting new people and so on. The worst thing we can do is to fill our lives with routine, repetition and familiarity. These make time pass so quickly that anniversaries and birthdays rotate faster every year, like a roundabout that picks up speed with every rotation. In fact, when people at my talks do disagree that time is speeding up as they get older, it's usually because they have recently been through eventful periods, such as living abroad or starting a new career.

Another perhaps more effective way of resisting the speeding up of time is to "de-automatize" our perceptions. This simply means practising conscious awareness or mindfulness, paying attention to our day-to-day experiences of seeing, hearing, feeling and so on. In the long term, we can cultivate conscious awareness through meditative practices that quieten the chatter of our minds and weaken the power of the conceptual labels that filter out the raw reality (or "suchness", to use a term from Zen Buddhism) of the world. These practices can also help us slow down time by softening our sense of separation from the world. (We will examine these methods in more detail in the last chapter of this book.)

However, it's certainly not easy to defy the first law of psychological time, as it is such a fundamental feature of human experience.

2. Time seems to go slowly when we're exposed to new environments and experiences (or inversely, time goes quickly when we're in familiar environments and have routine experiences).

This law is the basis of the first strategy of slowing down time I mentioned above, and of William James's point that "rapid and interesting" travel has a time-expanding effect. I mentioned my own examples in the Introduction (see page 4), when I went to live in Germany and then travelled around India. As I recounted in *Making Time*, I once did a survey at an airport, asking returning travellers what type of trip they had been on and whether they felt their time away had passed quickly or slowly. I found that people who had been on more adventurous trips – such as backpacking holidays, trips to non-European countries or working trips – reported a slower passage of time than people who had been on package holidays in resorts.

This was presumably because the former experienced more newness and unfamiliarity. In the latter case, people often mentioned that the first few days of the holiday passed quite slowly, but the second week had "flown by". This was presumably because they stayed in the same environment, which quickly became familiar to them. Dinah Avni-Babad and Ilana Ritov conducted a similar study in which 41 people were interviewed at the end of their vacations. The vast majority reported that the first few days passed quite slowly, before time started to speed up.[26]

This law of psychological time is also due to information processing. When environments and experiences are unfamiliar, we pay much more attention to them and so process much more information. In fact, this is one of the reasons why vacations are so enjoyable and so essential. They enable us to rekindle the fresh, "first-time" perception of young children, which stretches time. On the other hand, when we're surrounded by familiarity, we often switch off to our surroundings. We don't pay conscious attention to our environment and our experiences and so process little information. Whereas on vacation we're surrounded by freshness and newness, at home our experiences may feel mundane and stale.

As noted above, William James felt that this was an effect of memory. Time stretches because of the increased number of memories that new experiences create. However, I don't think this wholly explains the phenomenon. As increased information processing takes place in the present, it seems logical to assume that the time-slowing effect must be a present tense experience too. It's tricky to ascertain, because we can only judge time retrospectively when we look back at the end of a day or a vacation. But this doesn't necessarily negate the possibility that time expansion (or contraction) takes place in the present. *All* human experience takes place in the present, including our varied experiences of time. (In the next chapter, I will argue that TEEs are present-tense experiences too, rather than a "retrospective illusion".)

As with the first law of psychological time, it's not inevitable that familiar experiences and environments shrink time. In fact, as I described at the end of the last section (see page 21), one of the main aims of meditation and mindfulness is to transcend our normal automatic perception and perceive the world with fresh, unfiltered awareness. This is perhaps one reason why many spiritual teachers and traditions compare spiritual awakening to returning to childhood. For example, the ancient Taoist text, the *Tao Te Ching*, advises us to "return to the state of the infant".[27] In Taoism, the aim of practices like Tai Chi and Chi Gong is to become as physically supple and flexible as children. And this applies mentally too. The adult mind should become as spontaneous and open as a child's, with the same freshness and sense of wonder.

3. Time seems to speed up in states of absorption
(or inversely, time seems to slow down in states of non-absorption – for example, boredom or impatience).

Has time flown by while you've been reading this chapter? If not, it's either a sign that my writing isn't entertaining enough or that you're in a tired and distracted mood.

Another informal experiment I often conduct at workshops or seminars is to divide the class or audience into two groups, giving them a different text to read. One group receives a well-written, entertaining article about time from a popular magazine, while the other group attempts to read a heavy, academic paper about time perception, full of scientific jargon and statistics. At the end of the exercise, I ask the groups to estimate how much time has passed. The second group always estimates more time than the first, presumably because they were in a state of boredom. The first group – with the entertaining article – slip easily into a state of absorption, which has a time-contracting effect.

Consider what happens when you're playing a game of chess, playing the guitar, dancing or chatting with friends in a bar or café. In these situations, we often experience what the psychologist Mihalyi Csikszentmihalyi calls "flow" – a state of intense absorption in a stimulating or challenging activity. Flow is an essential component of a life of wellbeing. It makes us feel mentally alert and energetic, with a sense of control over our minds and a sense of inner harmony. If there is any negative aspect to flow, it is that it makes time pass very quickly. In flow, we don't usually have any awareness of time, because our attention is wholly absorbed in the activity. But when we emerge from flow and check our phones or watches, we're often astonished at how much time has passed. As Csikszentmihalyi has put it, in flow "Often hours seem to pass by in minutes; in general, most people report that time seems to pass much faster."[28]

Flow is not the same as mindfulness. It could be seen as a very narrow form of mindfulness – that is, mindfulness of a specific environment or activity. But in flow, we're unaware of everything *outside* that particular situation. Our attention is like a spotlight that illuminates a small area in great detail. In mindfulness, our awareness is more diffuse, like a powerful lamp that lights up a wide area. It is open to the whole landscape of our experience and environment, rather than deeply absorbed in a particular location. This difference is indicated by the fact that while flow speeds up time, mindfulness tends to expand it.

Absorption is the root of the saying that "time flies when you're having fun." Fun always involves absorption. When activities are enjoyable, we give our whole attention to them. We lose ourselves in them, forgetting our surroundings and ourselves – and the concept of time. As a result, absorption usually induces a time *contraction* experience. Note, however, that this is a relatively *mild* effect. Even at its most extreme – for example, when a deeply absorbed musician or computer gamer experiences seven hours as if they are two or three – it doesn't compare in magnitude to the drastic time expansion that often occurs in accidents or psychedelic experiences.

From the other perspective, when our attention isn't absorbed in an activity or entertainment, then time drags. This is essentially what boredom is: a state of non-absorption, in which we have nowhere to focus our attention and are left alone with our own thoughts. This is why we find waiting rooms such awful places, and why time drags so much in traffic jams or when a train is delayed. In these situations, it's difficult for us enter a state of absorption, as we feel distracted and agitated and often don't have access to entertainment and activities. We are acutely aware of the passing of time, which seems painfully slow.

This also explains why time drags in negative states such as pain, anxiety or depression. In boredom, our minds are unoccupied either because we find it hard to concentrate or because we can't find an activity or entertainment that seems interesting enough to focus on. In states such as pain or anxiety, we don't have the mental resources – or the inclination – to focus our attention on activities or entertainment. Our negative feelings are so strong that we can't focus away from them. We can't enter states of absorption and so remain acutely conscious of the passage of time. As one chronic pain sufferer told Ruth Ogden and her team, "Time goes a lot slower because I am unable to lose myself in a task that *would* distract me. Pain keeps my focus on exactly that and that alone."[29]

In my view, the relationship between time and absorption is also due to information processing. In absorption, we process comparatively little information. We obviously process information from the activity that engages our attention (for example, a book or a film or a game of chess) but this is quite a small amount compared to other states of mind. We narrow our attention to one small focus and block out all other potential sources of information in our environment. Most significantly, our minds become quiet – largely free of thought – so that we process very little cognitive information.

However, when our attention is diffuse or unfocused – in such states as boredom, impatience or anxiety – a massive amount of cognitive information flows through our minds. They fill up with incessant "thought-chatter" – thoughts about the future or past, fragments of conversations or songs, thoughts about politics or celebrities or pieces of news, daydreams in which we fulfill our desires, and so on. Hundreds, if not thousands, of such thoughts may pass through our minds in a matter of minutes. And all this cognitive information stretches time.

It's important to note that the above three laws of psychological time often operate concurrently, sometimes counteracting or intensifying each other. As I noted near the beginning of this chapter (see page 11), when people listen to uneventful ambient music for a long period, the time-contracting effect of the lack of information is usually counteracted by the time-expanding effect of boredom. In a similar way, as I also suggested above (see page 21), the first law (that time speeds up as we get older) can be mitigated by the second if a person undergoes a lot of new experiences, such as travelling around the world or starting a new career or relationship. Alternately, the first law may be intensified if a person lives a very uneventful life, repeating the same experiences in the same environment for decades. The first law may also be intensified by prolonged periods of absorption – for example, if you spend hours a day watching television.

4. Time passes very slowly in intense altered states of consciousness, when our normal psychological structures and processes are significantly disrupted, and our normal "self-system" dissolves.

Hopefully, I've made it clear that the first three laws of psychological time are closely related, in that they can be explained largely in terms of information processing. However, the fourth law of psychological time stands apart from the first three. Firstly, it *isn't* related to information processing (at least, not primarily) but due to other factors. Secondly, the fourth law brings a much more radical disruption to our normal time experience. The second and third laws may cause time expansion in a limited sense, such as when we travel to unfamiliar places, or (in a negative sense) when we're stuck in waiting rooms. You could compare these changes to a train that sometimes slows down or speeds up on its journey along the track. However, the fourth law of psychological time describes experiences of a completely different order of magnitude. To extend the last analogy, it's like stepping off the train altogether and finding ourselves in a strange, panoramic landscape, which we can only walk through very slowly. It's like stepping into a completely different "timeworld" where the normal rules of duration don't apply.

By the way, you may not be familiar with the term "self-system." It refers to our normal sense of identity and all the normal psychological processes and functions that constitute it. When these processes and functions operate normally, they create our normal consciousness and our normal awareness of the world. When it is disrupted, we experience altered states of consciousness, which means – among other changes – entering a different timeworld.

And it's this strange and exhilarating timeworld that we're going to step into now and spend the rest of the book exploring.

CHAPTER 2
IN THE
DANGER ZONE

Time Expansion in Accidents and Emergencies

Of all the situations in which time expansion experiences occur, the most common is in accidents and emergencies. In my research, I have found that just over half of all TEEs occur in this context.

Emergency TEEs have several core characteristics, along with their fundamental feature of time expansion. For example, they usually involve a sense of wellbeing. Even though their lives might be in danger, people feel calm and relaxed. They may even describe their experience in "spiritual" terms, with characteristics of serenity, transcendence and oneness. Emergency TEEs (or ETEEs, as I will refer to them from now on) also usually involve a sense of alertness or heightened awareness. Perception becomes much more vivid and intense than normal, so that people notice more detail and beauty.

The above characteristics are common to TEEs generally, as we will see later. But there are some characteristics that are particular to ETEEs (and also to sporting TEEs, which we will examine in the next chapter). One is very rapid thought and action. In accident situations, people are often surprised by the amount of time they have to think and act. Another characteristic, particularly in life-threatening situations, is a sense of detachment, as if the person is watching from above or from a distance. Less frequently, there is a sense of external sounds becoming muffled or wider surroundings being blurred.

In this chapter, we're going to examine ETEEs in detail, beginning with falls.

TEEs in Mountaineers

I'm not the first researcher to systematically study time expansion experiences. In fact, you might be surprised to learn that the first study of TEEs was published as long ago as 1892, although not by a psychologist but a geologist.

Albert Heim was a Swiss geologist and climber, born in 1849. In 1871, he was leading a climbing party up a mountain in eastern Switzerland when he lost his footing and slid over a cliff, falling around 20 metres (66ft). According to the laws of physics, a fall of 20 metres (66ft) lasts around two seconds. But to Heim, those two seconds stretched infinitely longer. He felt calm, completely free of anxiety, with clear and coherent thoughts. He had ample time to deliberate over how he should respond to his predicament. He wondered whether he should keep hold of his alpenstock (a long walking stick with a metal point) and whether he should take off his glasses. He decided that if he survived the fall he would immediately shout out to his companions, so that they wouldn't endanger themselves climbing down to search for him.

Besides these practical points, Heim pondered over a variety of other issues. As a PhD student, he was scheduled to deliver his inaugural university lecture in five days. He realized that now – assuming he was going to die – this would be impossible. He imagined his relatives and friends receiving the news of his death and felt compassion for them.

Then, perhaps most strikingly of all, Heim saw a review of his life. Living before the advent of cinema, he used the analogy of a play to describe the experience:

I saw my whole past life take place in many images, as though on a stage at some distance from me. I saw myself as the chief character in the performance. Everything was transfigured as though by a heavenly light and everything was beautiful without grief, without anxiety and without pain. The memory of very tragic experiences I had had was clear but not saddening . . . Elevated and harmonious thoughts dominated and united the individual images and, like magnificent music, a divine calm swept through my soul.[1]

After the review, Heim watched himself falling through the air, with a snow-laden field beneath him. He heard a dull thud as he hit the ground, then lost consciousness. It was half an hour before his companions found him and carried him to the nearest Alpine hut. Miraculously, he escaped serious injury and was able to deliver his inaugural lecture the following week.

Over a decade later, now a professor of geology at Zurich Polytechnic, Heim contemplated his fall again and decided to find out how common such experiences are. He collected 30 accounts of near-fatal falls, not only from climbers but also from soldiers, construction workers and others. In 1892 – just two years after James's *Principles of Psychology* was published – he presented his findings to the Swiss Alpine Society. He also published them as a paper entitled "Notes on Fatal Falls".

All the reports featured essentially the same characteristics. The fallers felt no anxiety or pain, but a sense of calm acceptance and mental clarity. Most strikingly, "Mental activity became enormous, rising to a hundred-fold velocity or intensity . . . Time became greatly expanded. The individual acted with lightning-quickness in accord with accurate judgment of his situation. In many cases there followed a sudden review of the individual's entire past."[2] Many people felt, like Heim, that their massively slowed down sense of time and rapid thought processes enabled them to take preventative action, and even helped them to survive.

Is it possible that Heim was embellishing the cases to conform to his own experience? This seems unlikely, as there have been so many independent reports of similar experiences from other climbers. Even while Heim was alive, an Austrian climber called Eugen Guido Lammer reported a similar experience, which Heim was apparently unaware of. Lammer was one of the most famous and daring climbers of the late 19th century, known as "the fetterless", as he would climb dangerous routes without guides and with minimal equipment. In 1887, he was attempting to climb the treacherous western face of the Matterhorn with a companion, when an avalanche suddenly descended and swept them away. As Lammer fell down the side of the mountain, he felt that he "hovered above this entire incident as a quiet, curious spectator". While observing the fall, his mind flooded with thoughts and memories. As he put it, "I would have to fill hundreds of pages with this mass of images and ideas . . . Years passed during the fall, centuries." Falling around 200 metres (656ft), Lammer was certain he was going to die, though unconcerned at the prospect. But somehow he survived. As he wrote, "Then the avalanche's waterfall-like roar became quieter . . . I opened my eyes and endless amazement came over me."[3]

One of the leading mountaineers of the second half of the 20th century, Doug Scott, reported a similar experience. A specialist in high-altitude and "big-wall" (large vertical expanses of rock) climbing, Scott was the first English climber to reach the summit of Mount Everest, in 1975. In 1992, he was caught in an avalanche while climbing the Mazeno Ridge in the Himalayas. Tumbling down a massive 500-metre (1,640-ft) ravine, he described how time became suspended, and he "found myself observing everything I experienced, as though from a bubble". He felt no fear or pain and was amazed by the resilience of his body. He compared his slow and serene movements to floating in a hot air balloon.[4]

Other Types of Falls

Very few of us have experienced spectacular falls from snow-laden peaks, but perhaps you've had a TEE during a more mundane type of fall. Any type of fall may give rise to a TEE, no matter how long in terms of distance or time. This includes short falls from a horse or a bike, a tumble down a few steps or simply losing our footing while walking.

Here I'll share some reports from my own research. In the following example, a man described his experience of falling downstairs:

> I hadn't zipped up the zipper of one of my boots properly . . . I ran down the stairs, lost balance and sailed down. During flight I suddenly felt like I was next to my physical self (parallel to my body) and in a state of slow motion. I was perfectly calm, thinking, "I must pull my knees up to my chest." The next thing I felt was me landing on the stairs, riding down the way a slide goes downhill. I was unharmed, only my shins and knees turned black and blue later. The most extraordinary experience. Probably saved me from breaking my skull or neck.

In the following account a man described an experience he had while he was waiting at a bus stop and witnessed the elderly woman next to him falling over:

> She fell sideways to her right. I thrust out my left arm to catch her. However, because I was off-balance, and she seemed to have lost consciousness and become a dead weight, I was pulled over with her. What followed was a remarkable piece of slow-motion choreography allowing me time to twist my body around her in such a way that I fell first with her on top of me. I also had time to completely relax my body so that I was not injured in

any way when I hit the pavement. All of this happened in what must have been a split second in real time.

"Choreography" is the best way of describing this, I think, because it seemed, looking back, that I was in a different space observing and issuing instructions to my body. And somehow I was granted the time to do everything!

In a final, more dramatic example from my research, a woman called Catherine detailed a TEE that occurred when she fell off a horse:

The whole experience seemed to last for minutes. I was ultra-calm, unconcerned that the horse still hadn't recovered its balance and quite possibly could fall on top of me . . . My thoughts were only, "I wonder where the horse is?" When the horse did pass me and knocked my head, I was still unconcerned and didn't feel a thing then or when I hit the ground.

The sound was muffled. I could see the other rider who was watching, but he was blurred. I was engrossed and amazed at the detail in the grass I could see while on the ground. I could see individual blades of grass and the intensity of colour was beautiful. Eventually I was reorientated to present time. That's when I started shaking and feeling the pain.

It's clear that these more mundane falls contain most of the elements of mountain falls. As well as massive time expansion, there is the same calmness, clarity and detachment, together with a feeling of being able to take preventative action. The only element that mundane falls don't normally feature is the life review, perhaps because they don't usually involve the same life-threatening danger as mountain falls (although the fall from the horse above was potentially life-threatening). As we'll see in Chapter 6, the life review is almost always associated with life-threatening danger and also with situations when a person *does* actually clinically die for a short time, before they are resuscitated.

TEEs in Traffic Accidents

Until a few decades ago, falls were probably the most common source of TEEs. However, now the most common source by far is traffic accidents – mostly cars and motorcycles, but also bicycles. My research suggests that these constitute around almost three-quarters of ETEEs and just over a third of *all* TEEs. This reflects how dominant cars have become in modern life, with the ever-present danger of accidents.

I have collected so many accounts of traffic-related TEEs that it's difficult to know which ones to use as examples. Let's begin with a couple of typical ones. Here a woman describes a TEE that occurred when she was a passenger in her friend's car, heading down a steep hill beside a river. As she recalls:

> It was winter and very icy. There was a steep drop of at least 30 metres (100ft) with only a tiny guard rail. We went into a skid and the car started spinning toward the guard rail. Time slowed way down.
>
> I was completely calm and had plenty of time to think or react but as I wasn't driving there was nothing I could do. It seemed like I was super alert and very relaxed and calm. We were spinning in super slow motion and the funny thing was the sound of the passenger in the front seat screaming. It sounded like a record played at the wrong speed . . . Luckily, we spun around a few times and the guard rail stopped us. No one was hurt but the car . . . I had all the time in the world.
>
> Even though it happened so long ago the memory is still very vivid in my mind. There was absolutely no fear, just a kind of hyperalert calmness.

The next report is from a motorcyclist who was travelling at around 30 mph (48 kph). As he approached an intersection, a car turned onto the road in front of him, failing to see his bike:

> It seemed certain that I would hit him. Then something happened. I felt like I woke up, like when you suddenly wake up in the night. Everything stopped. I felt really calm. I looked at all my options and decided to go the opposite way, even though there was another car ahead. I knew that I could go around that car if I used the empty sidewalk.
>
> It was a really cool feeling. I watched my bike sliding in slow motion and hoped it wouldn't start tumbling because the damage would be far more expensive. I looked ahead and thought, "At least I'll stop before I hit the mud puddle 20 yards [18 metres] away in the grass." Next thing I knew my face shield was covered in muddy water. I had slid right through it and was still moving in the grass.

Both the above experiences are similar to my own TEE, as described in the Introduction (see page 1). In fact, most traffic-related TEEs are remarkably uniform, featuring the characteristics of calmness, alertness, rapid thought and the capacity to take preventative action. However, one variation is that some more severe accidents – as with especially serious falls – feature a life review. For example, a man described a TEE that occurred when his friend was driving at high speed down a narrow country lane. His friend lost control of the car and clipped a grass bank. The car flipped and spun through the air: "It was like I recalled every experience I'd ever had at once, all memories and feelings and emotions. It included all kinds of things that I'd completely forgotten about."

Sometimes traffic-related TEEs are shared by passengers. One woman describes how she was driving, with her daughter in the passenger seat, when their car was hit by a van: "We both experienced the slowing down experience and a calmness. The windscreen shattered in slow motion and neither of us experienced any panic until we came out of this sort of trance." Another person reported a TEE he shared with his wife when a large four-wheel drive went through a red light and hit their car:

> We went way into the air, rotating end over end and sideways. We witnessed everything in slow motion and described it in exactly the same way afterwards. The amount of detail we recalled as this massive four-wheel drive arched high above our car, including its rotational patterns, was absolutely staggering. We both clearly recall an eerie silence, calmness. No sound whatsoever. Then our car landed on its side about ten metres [33ft] away with a bang and normal time/speed reengaged.

However, TEEs aren't normally shared experiences. For example, in the first traffic accident above, the passenger who was screaming presumably didn't have a TEE (since TEEs also involve a sense of calmness and lack of anxiety). In my own accident, my wife was sitting next to me, but didn't share my TEE.

Why should some passengers share TEEs, while others don't? This is probably due to the link between TEEs and altered states of consciousness. The sudden shock of an accident may disrupt our normal psychological processes and functions, causing an abrupt shift in consciousness. This is vividly captured by the motorcyclist above, who said, "I felt like I woke up, like when you suddenly wake up in the night." So, TEEs are presumably shared when two or more people shift into an altered state of consciousness in the same situation, which may not always be the case. Some people may be generally less susceptible to altered states or perhaps their psychological state was less amenable to a shift at that moment.

In a more general sense, this helps to explain why not all accidents and emergencies result in TEEs. It depends simply on whether a person shifts into an intense altered state of consciousness or not. This, in turn, depends on two things: how susceptible a person is to altered states and how severe or shocking the incident is. The more severe and shocking it is, the more likely it is to induce an intense altered state.

This leads to another important question: how common are TEEs in accidents and emergencies? So far I haven't investigated this question specifically but there are some helpful pointers from earlier research. In the 1970s, two psychologists at the University of Iowa, Russell Noyes and Roy Kletti, studied over 200 examples of life-threatening accidents, and found that around three-quarters featured dramatic time-slowing. Admittedly, this research focuses on the more extreme side of the spectrum – life-threatening accidents – but it does indicate that TEEs are fairly common. (Incidentally, in line with my own findings, many of Noyes and Kletti's participants reported heightened attention and alertness, and the ability to respond quickly and effectively to their predicament. Around a third of participants reported extremely rapid thought processes, up to 100 times faster than normal.[5])

Sense of Beauty

One unexpected aspect of traffic-related TEEs is the sense of *beauty* that people sometimes describe. People who report TEEs during falls often describe a sense of beauty, as with Albert Heim (see page 31), or the woman who fell off her horse (see page 34) who reported that "I could see individual blades of grass and the intensity of colour was beautiful." In view of the ugly destructiveness that we associate with traffic accidents, it's perhaps surprising that some traffic-related TEEs feature similar descriptions. For example, one man who had a car crash told me how "the glass shatter[ed] beautifully in front of me in slow motion, from bottom left to top right like a river branching out." Another person gave a similar but more vivid description of his perceptions after being knocked off his motorbike at high speed:

> I saw the car's windscreen shattering. The glass sprayed out so slowly, like a fan, and it looked beautiful. All the pieces were shining in the sun. I felt like I was floating through the air, almost as if I wasn't going to come down. I looked into the sun and it was like being on a plane, when you're above the clouds and it's a brilliant white colour.

These quotes also illustrate the heightened awareness that often comes with TEEs. To some degree, heightened awareness is caused simply by slowness of perception, which enables people to perceive more detail in the same way that they can think in more detail. But it's also important to note that heightened awareness is a feature of many intense altered states of consciousness. We take our normal state of consciousness for granted, assuming it provides a reliable and objective experience of reality. However, some altered states suggest that our normal awareness of the world is limited and diluted, like an old black and white photograph compared to a modern virtual reality experience. Our normal awareness seems to be filtered and diluted by familiarity. But in some altered states of consciousness – including TEEs – we transcend this limited vision and "wake up" to a brighter, clearer and more beautiful world.

Other Types of Emergency TEEs

The incidents that induce TEEs have three essential qualities: they must be *unexpected*, *sudden* and *dramatic*. If we're told in advance that an event might happen, or if it occurs gradually or if it isn't extraordinary or unusual in some way, then it won't have the same power to disrupt our normal consciousness. In other words, ETEEs must involve shock and potential danger. This also means that they cannot usually be *consciously* induced. If we prepare for them or anticipate them, it's highly unlikely that they will occur.

Beyond traffic accidents and falls, almost any sudden emergency can jolt us out of normal consciousness into a different timeworld. This includes fights, assaults, explosions, health emergencies and natural disasters such as earthquakes. For example, a woman described a serious asthma attack in which she "became timeless with a distinct sense of calm despite the seriousness of the situation". A man described an incident in which he was stung three times inside his mouth by a wasp: "Time was slow but also did not exist. As ridiculous as it sounds, it felt a bit like the film *The Matrix*, where the scene slowed right down then stopped." Similarly, an American policeman recalled how "time slowed to a crawl" when he came face to face with a gunman. A soldier reported an incident when his group was under mortar attack and time slowed down drastically, enabling him to "see the shock waves and shrapnel as they exploded, which mostly happens too quickly to be of any note . . . Even the return rocket seemed to glide slowly through the sky, quite uncharacteristic of a rocket."

Here is one of the most dramatic TEEs in my collection, from a woman who saved her children from a fire. This is a good example of how TEEs facilitate preventative action that would be impossible in normal time, which helps the person experiencing it – and, in this case, also the woman's children – to avoid danger and potential death:

> I woke up to see flames coming from the empty bus next to the one we were all sleeping on and then went about getting my children (who were very young) out and away from danger. I will never forget the moments

of absolute clarity and calmness. It didn't feel like I was even in my own body. I first moved one child out and handed her over to the girl who came to help, and then I went back and woke up my eldest, scooped up the baby, took my eldest's hand and we got off.

The details are still with me: the sight, sound, smell and temperature of the fire; the face of the girl who helped, lit up from the reflected fire; and the sight of the sleeping kids.

Thankfully we all got away safely. I think the only reason I was able to do this was that I first experienced a great calmness and time seemed to stop.

In another dramatic example, a woman described a shooting that occurred when she was four-and-a-half months pregnant. Two gunmen arrived at her house as her husband was leaving for work. In her words:

I ran to the window and saw him with these two men jerking back and forth . . . We were living in a two-storey structure. I was on the upper storey. There was a balcony in the bedroom. I looked out and could see that they were trying to force open the sliding glass doors to the living room . . . My husband had given me a blank gun and I took it out of the drawer . . . I stepped out on to the terrace, and I was just going to fire a couple of shots so the neighbours would call the police. At this point the two men had given up trying to break in and they were dragging my husband to the main door. We were probably 30 feet [9 metres] apart, me above and them below. I just pointed a gun and then realized that one of them was pointing a gun at me. I heard a click sound and it was his bullet hitting the concrete wall maybe a foot away from my head.

My first reaction was to laugh, and I remember saying to myself "Wow, he's got a real gun." This man

41

was shooting at me, three more times. It shattered the windows. The night before the attack there was a movie on TV with a French actor who played an Israeli policeman who throws himself to the ground to avoid being shot. I remembered it and decided I should do the same. I turned around and everything slowed down enough for me to think: "I have to get to the ground, but I have to fall in such a way as not to damage the baby in my belly." There was no fear – I knew what I had to do. I saved myself and my baby. The four shots probably happened over two seconds, but it felt like about 15 seconds to me. Time totally slowed down and allowed me to think.

In a slightly different variant, ETEEs may also occur in moments of intense psychological turmoil, during traumatic life events or following tragic news. We sometimes use the phrase "time stood still" in these situations and this isn't simply a figure of speech. Trauma and turmoil can stun our minds in the same way as accidents and emergencies, shifting us into an altered state of consciousness.

I received reports of this type of TEE from situations such as bereavement, divorce and the diagnosis of a serious illness. One woman told me how "time did really feel as though it had stopped" when she received a phone call with the news of her husband's death. Another woman described a TEE that occurred when she made the decision to separate from her husband. As the certainty of the decision took hold in her mind, she "experienced [a] sudden burst of clarity and energy expansion, I can't really express it well in words, but from the grief over my loss, time stood still for a moment . . . it created such an intense sense of presence in me, and peace."

Finally, a woman described a TEE that occurred while she was awaiting the results of a test for breast cancer. She told me, "I intuitively knew what the outcome of the tests were and as I sat in the waiting room everything slowed down . . . the people walking on the sidewalk outside the window and everything in the office as well. The smells, sounds and sights were very vivid."

Preventative Action

Again and again in ETEEs, people report that they have abundant time to think methodically and practically about how to respond to their predicament. They describe feeling that they have enough time to anticipate and avoid danger or to position themselves to minimize injury. As one person who was involved in a car crash put it, "I will always remember how much time I seemed to have to think and work things out." Or as the woman in the shooting described above reported, "Time totally slowed down and allowed me to think." Many people were certain that this extra reaction time had saved their lives – sometimes other people's lives too. In the above example of the woman who saved her children from a fire (see page 40), she said she was only able to do this because "I first experienced a great calmness and then . . . time seemed to stop."

In my research, just over half of all ETEEs feature this capacity to take preventative or protective action. However, it's important to note that many emergency situations don't allow for preventative action – for example, when a person is a passenger in a car, or in health emergencies such as (in the above examples) an asthma attack or a wasp sting. This obviously also applies to trauma-related ETEEs such as bereavement or a diagnosis of cancer. Setting these examples to one side, my research suggests that *in almost every situation where it is possible*, people report the ability to take preventative action.

Certainly, this applies to the great majority of traffic accidents. For example, here a man described how he survived a car crash in which another passenger died:

In the split second between loss of control and impact . . . somehow time slowed so much that I was able to crawl down and curl up into a ball under the glove compartment on the front floorboard, hoping the confined structure of floor, dash and front of seat would help prevent injury. Upon impact the car flipped end over end several times for a distance of about 50–75 yards [46–69 metres]. The car was crumpled into an unrecognizable ball of twisted metal. I survived with only minor injuries.

A woman described an accident when she was knocked off her bicycle by a spray of water from a water truck, then run over by a gravel truck. Like the car passenger above, she was convinced that time expansion enabled her to take protective action that saved her life. This example is also interesting because it includes estimates of actual and subjective duration:

When my lawyers asked me how long [it was] between the time I was first hit with water to the time I was hit [by the gravel truck], my best estimate was 3.5 seconds. But those 3.5 seconds felt like 40. I had lost my balance and I managed to push myself off the gravel truck with my hand but then I ricocheted off the water truck and went down just in front of the last set of double wheels. I still had, in those 3.5 seconds, enough time to figure that the wheels were aimed at my abdomen, and I pulled myself further under the truck before the wheels hit, so they hit my pelvis and my muscular thighs, and I survived. I know I definitely wouldn't have if the wheels had not hit the strongest part of my body.

Interestingly, like Albert Heim (see page 30), the woman described some of the detailed thoughts and observations that streamed through her mind during those three-and-a-half seconds:

> Before the tires hit, I was preparing myself for being a paraplegic and possibly having to navigate college in a wheelchair. I pictured myself in a wheelchair at the local university dances. (Becoming a paraplegic would be the only way I went to the local university instead of leaving the state.) My folks were out of town, and I had time to consider that was pretty typical – one of their children only went to the ER when they were out of town. I thought that for sure my mom would fly up – at that point in time the only way my mom would be in the same room with my dad was if it was one of their children's hospital rooms . . . I had all that time to think of that and after I hit the ground and before the tires hit me.

The feeling of being able to take preventative action occurs in most falls too, despite their short duration. Physics tells us that a fall of two metres (6½ft) lasts just over half a second, while a fall of five metres (16ft) lasts one second and a fall of 20 metres (66ft) lasts around two seconds. In other words, when you lose your footing and fall to the ground, it lasts around half a second, while if you fall from the roof of an average two-storey house, it lasts around one second. During these brief periods, people often report the ability to make complex movements, along with detailed thought processes. A man described a TEE that happened while he was working out by running up and down staircases at a stadium and lost his footing about four steps from the bottom. Interestingly, this experience was witnessed by a friend, who was also surprised by his rapid protective action:

> I was tripping, falling headfirst to the hard steel staircase. Then everything starts to go in slow motion. I manage to

process everything, literally 12 inches away from hitting the floor headfirst . . . I start to move my legs and body in a weird way to where I can push myself upward so I don't fall on my head. My friend said, "I was certain you were going to fall headfirst." I told him everything went into slow motion, but he didn't really believe me.

Similarly, an elderly man described a fall while out walking his dog. Suddenly, the dog walked in front of him, throwing him off balance. According to the above calculations, this fall would have lasted no more than half a second:

As I fell forward, I was aware of calculating how best to fall so as not to have too great an impact on any one part of my body. As my knee brushed the ground, I was conscious that reaching out would shift the force of impact. The ground was stony, but my hands were protected by thick gloves, and I noted that as I fell. I also had a sense of self-congratulation as I became aware of how I had shifted the point of impact. I lay still for a moment mentally checking my body. I felt quite elated about the experience, marvelling at the reaction of my mind and body . . . I sustained no injuries and carried on walking.

Is it possible that this kind of preventative action could occur in normal time? After all, human reflexes operate rapidly in all circumstances. Why shouldn't a person be able to adjust their posture – to shift around their body or move their limbs – in the space of a half a second? Perhaps people retrospectively attach thought processes to their reactions, while at the time they simply act instinctively, without deliberation.

It's difficult to judge this precisely, but I would argue that in most of the above examples, the complexity and intricacy of people's movements far exceeds what would normally be possible within such brief periods of time. The anthropologist

Edward T Hall provides a striking example of this in his book *The Dance of Life*, where he reports the case of a pilot who faced a crash shortly after take-off. As soon as the pilot realized he was in danger, time slowed down drastically. In the next eight seconds, he managed to avert disaster, undertaking a series of actions and manoeuvres so detailed and complex that they took over 45 minutes to describe.[6]

It's also important to note that, in many cases, people's movements are clearly conscious and deliberate, rather than instinctive – for example, when the pregnant woman consciously dived to the ground in a way that would avoid harming her baby (see page 42), or when the woman who was knocked off her bike by a water truck had time to "figure" that the wheels were heading for her abdomen and pulled herself further under the truck to avoid this. In the earlier example of the man who fell at a bus stop while trying to stop an elderly lady falling (see page 33), he described being "in a different space observing and issuing instructions to my body". Another man whose car went into a spin on a motorway recalled consciously giving himself instructions, which he recalled as "Don't touch the steering wheel – I'm sliding in the right direction, toward the hard shoulder! Look back up the motorway, is anything coming toward me? No – okay, if there is I have time to abandon ship." Finally, a man described a TEE when a metal barrier was falling toward his car. As he told me, "My mind was telling me, 'move the car to the right, some more, some more – move.' I did and was able to miss the metal barrier."

TEEs Are Real Experiences

In other words, the sense that most people have of being able to take preventative action in ETEEs is real. And in turn, this shows that ETEEs themselves are real. A sceptical interpretation might be that they are hallucinations caused by unusual neurological activity in states of extreme stress. (We will discuss some other attempts to explain TEEs and TCEs in terms of unusual neurological activity in Chapter 7, once we have completed our survey of them.) However, as people really do have increased time to think and act, this can't be the case. This applies to perception too: because they actually do experience more time, people are able to perceive the events and their surroundings in much more detail, in the same way that if you walk slowly through a landscape, you perceive much more detail than if you are running or driving.

The clarity and heightened awareness of TEEs also strongly suggests that they are real experiences. When hallucinations are over or when we wake up from dreams, we have a clear sense that we've returned to reality from a delusory state. But after TEEs, we usually have the opposite sense – that the experience was *more real* than our normal state. For example, a prison officer reported a TEE that occurred when he was assaulted by an inmate: "Time slowed down in an instant. My hearing was such that every noise sounded muffled . . . I could feel my body in slow motion." He described the experience as "superreal".

It's also significant that occasionally people report hearing voices at a slower speed too, such as the car crash victim who heard the scream of another passenger as "like a record played at the wrong speed". In the above incident with the prison officer, the inmate's assault was witnessed by his boss, who shouted an obscenity that sounded like "listening to a tape at quarter speed. It actually sounded like that."

Admittedly, this slowing down of sound isn't a common feature of ETEEs. However, it's worth nothing that very few ETEEs actually feature speech. Apart from the fact that many of them happen to people who are alone, people are often too stunned to speak or are too busy trying to take preventative action. Of course, some ETEEs may feature other human noise, such as screams and shouts, as well environmental noise.

However, by far the most common auditory feature of TEEs is muffled noise, quietness or silence. Perhaps this explains why more people don't comment on a slowing down of sound. Clearly though, the fact that some people hear sounds in slow motion indicates that their TEEs are real.

Do TEEs Take Place in the Present Moment?

Another sceptical way of explaining TEEs – besides dismissing them as simple hallucinations – is to suggest that they are an effect of *recollection*, a type of false memory, rather than a real experience that happens in the present.

This relates to what is probably the most well-known experiment ever conducted on time expansion experiences. The experiment was designed by researchers at the University of Texas and Baylor College of Medicine in 2007 to test the theory that TEEs are a retrospective illusion. At an amusement park, participants made 31-metre (102-ft) freefall jumps while harnessed to a platform, with a safety net to catch them. They had a device strapped to their wrists containing flashing digits. The researchers reasoned that if time slowed down, participants would have extra time to process information and so would have an enhanced ability to read the digits. However, this didn't prove to be the case, suggesting that time didn't actually expand for the freefallers. However, the participants did retrospectively overestimate the length of their falls, by an average of 36 per cent.

From this, the researchers concluded that the time expansion reported during falls is an "illusion of remembering an emotionally salient event". They suggested that apparent time expansion during emergencies is due to the increased number of impressions and perceptions we absorb, which creates more memories. (This is sometimes referred to as the "hypermemory hypothesis".) In other words, we don't actually experience time expansion – it just seems that way when we recall emergency situations, because our awareness becomes heightened, allowing us to take in more detail. As the researchers put it, "time-slowing is a function of recollection, not perception: a richer encoding of memory may cause a salient event to appear, retrospectively, as though it lasted longer."[7]

However, there are some serious issues with this experiment and with the conclusions drawn by the researchers. Most significantly, the jumps didn't constitute a *real* emergency. Real emergencies are unexpected and involve real, life-threatening danger. In the experiment, the participants were attached to a platform and had a safety net. As noted above, ETEEs occur when events are *unexpected*, *sudden* and *dramatic*. None of these criteria apply to the experiment. The participants had advance knowledge of the nature of the activity (including that it didn't involve real life-threatening danger) and no doubt underwent lengthy preparations for their jumps. In other words, the freefalls were in no way comparable with accidental falls, traffic accidents or other emergencies.

In fact, it's clear that the experiment didn't induce fully fledged TEEs in the participants. They only overestimated the falls by 36 per cent, which is nowhere near the normal time-stretching effect of TEEs. We've seen that ETEEs are related to altered states of consciousness, so presumably because they lacked the authenticity of sudden accidents, the falls didn't induce altered states.

As we saw in Chapter 1, it's certainly true that an increased number of impressions and perceptions have a time-stretching effect. Increased information processing slows down time, which is the basis of the second law of psychological time ("Time seems to go slowly when we're exposed to new environments and experiences"). No doubt this was why the freefallers experienced a mild time expansion. However, the time expansion of genuine emergency situations goes way beyond the mild time-stretching effect of this experiment or of a trip abroad, a new career or a relationship. It is simply impossible for increased information processing to cause the types of radical time expansion we have examined in this chapter. Clearly, another factor must be involved – namely, an altered state of consciousness.

Let's reverse the causal connection suggested by the above researchers. In TEEs, it isn't intensified perception that expands time, but time expansion which allows for intensified perception. As noted above, people perceive more detail during TEEs precisely because time is moving so slowly – as if they are flying over a landscape in an air balloon over a landscape rather than in an airplane, able to view much more detail below.

It's therefore invalid to apply the findings of this study to genuine TEEs caused by unexpected, life-threatening accidents and emergencies. Beyond this, there are some more general reasons why TEEs can't be a retrospective illusion. Obviously, if people can accomplish feats that would be impossible in normal time, then TEEs have to happen in the present – and the same goes for rapid cognition and perception. As we'll see in the next chapter, these points make even more sense in relation to sporting TEEs, when time expansion gives athletes a massive advantage, as they have more time to react or to complete intricate movements.

Certainly, most of my participants were sure that they were experiencing time expansion *in the moment* rather than in retrospect. A few had heard about the theory that TEEs are caused by recollection and were dismissive of it. The woman who was knocked off her bicycle by a spray from a water truck (see page 44) rejected the idea as "poppycock . . . Absolutely the perception of time slows at the time one experiences the emergency, not just in memory playback." The person who avoided a metal barrier falling on to his car (see page 47) told me, "It was an event that happened in the present, it's not a recollection of memories or events. For me the slowing down of the moment made me escape and decide how to escape the falling metal on us."

My doubts about the "retrospective illusion" theory are shared by the neuroscientist Dean Buonomano. After summarizing the "hypermemory hypothesis" as a possible explanation for time expansion, Buonomano refers to his own TEE in a car accident, when he recalls saying to himself, "Wow, time really does slow down." Like most of my participants, he was sure that "I did perceive events unfolding in slow motion *during the accident*." As a result, he states that "the hypermemory hypothesis cannot fully account for the slow-motion effect."[8]

How Slowly Does Time Pass during Emergency TEEs?

Although there's no doubt that time expansion is real, it doesn't expand in a uniform way. In my research, there was a lot of variation in the degree of time expansion people experienced.

Most commonly, the degree of time expansion is to the order of 10 to 40 – that is, one second is perceived as equivalent to 10 to 40 seconds of normal time. For example, a woman whose car was hit by a falling tree – which smashed her windscreen – estimated that "the time for [the tree] to hit then clear my car would've been about a tenth of a second" but she experienced roughly 3–4 seconds of normal time. Similarly, one person estimated a four-second time span as a minute, while the person who was hit by a water truck (see page 44) suggested that 3.5 seconds extended to 40 seconds. Similarly, in my own ETEE, the incident probably lasted 3–4 seconds, while I would estimate an experience of 30–40 seconds of normal time. Certainly, in many of the above accidents, the complex manoeuvres and perceptual and cognitive details were more compatible with a minute or more, rather than the few seconds that actually passed.

In other cases, a few seconds stretches into minutes. In other words, there is a time expansion to the order of 60 or more. One person in a car accident stated that "it seemed as though this took minutes, but it all happened within a second or two." The person who fell off a horse reported that "the whole experience seemed to last for minutes." Similarly, a man who fell off a balcony noted that, "it seemed like minutes but would have been seconds."

In some rare TEEs, time expands even further. It becomes so stretched out that it scarcely seems to exist. For example, one person described a car accident that lasted around 20 seconds, but which felt like "years of my life passing by". Another person reported a car crash that "seemed to last forever but it obviously was only seconds". As we'll see later, this type of extreme expansion frequently occurs under the influence of psychedelics or in near-death experiences.

This naturally leads to the question: why is there so much variation in TEEs? Perhaps again, this is related to altered states of consciousness. There are degrees of altered states – some mild, some very intense. And the more intense an altered state is, the more time expansion it brings. As a general rule, the most severe and life-threatening accidents cause the most intensely altered states, bringing more intense time expansion. This explains why some of the mountain falls we examined at the beginning of this chapter feature such dramatic time expansion. (As we'll see later, it's because psychedelics and near-death experiences induce such radically altered states of consciousness that they also usually cause dramatic time expansion.)

Susceptibility to altered states is also a factor. If a person is highly susceptible to altered states, they may experience dramatic time expansion even in a non-serious accident. If a person has low susceptibility to altered states, they may only experience a mild time expansion (or none at all) in a life-threatening situation.

A Different Perspective

If TEEs were just hallucinations or retrospective illusions, we could brush them aside and ignore their implications. But as they are real experiences that happen in the moment, they have a great deal of significance. On the one hand, they graphically emphasize that human beings' experience of time is highly flexible and subjective. They show that our normal experience of time is simply a manifestation of our normal state of consciousness. When our normal consciousness is significantly disrupted, we enter a different timeworld.

TEEs also point to a different view of reality itself. Like other altered states of consciousness, TEEs suggest that our normal awareness of reality is not – as with our normal time perception – reliable and objective. We learn that there is not simply *one* reality. "Normal" reality is simply one perspective, produced by our normal state of consciousness. And TEEs suggest that this normal perspective is limited. They seem to offer a glimpse of a brighter, clearer and more beautiful world, which appears more authentic than normal reality.

However, perhaps most importantly, TEEs bring a different view of *ourselves* and of our potential as human beings. In many of the reports I received, people were amazed at how calm and collected they felt, completely without fear or panic. Many people were also amazed at how effectively they acted, with a resilience and skill they never realized they possessed. In the aftermath of their TEEs, they reported a greater sense of confidence and self-esteem, now that they were aware of their inner strength.

This reinforces my long-held feeling that we human beings tend to underestimate ourselves. Some of us struggle through our everyday lives, feeling timid and awkward and overwhelmed by the world's demands. We feel inferior to others and suffer "imposter syndrome" in our professions, due to a deep-rooted sense that we're not "good enough". But TEEs remind us that there are deep reserves of inner strength inside us. When we face crises and emergencies, we awaken our highest selves and respond with a deep intuitive intelligence and resilience.

Finally, it's significant that ETEEs are usually experiences of wellbeing. In this chapter, we've seen many examples of people who faced imminent death with serenity, rather than panic and fear. In some life-threatening ETEEs, people feel such intense peace and joy that they lose all fear of death. So even though we might dread the thought of accidents, they may be paradoxically pleasant experiences that lead to profound and positive change.

CHAPTER 3
THE ZONE OF PEAK PERFORMANCE

Time Expansion in Sport

At the highest levels of sporting competition, most athletes are familiar with time expansion experiences. In fact, it is their familiarity with TEEs that – at least in many cases – enables them to *reach* the highest levels of competition.

You may have heard athletes speaking of being "in the zone", when they attain peak performance. Their attention becomes completely absorbed in the game or event, to the point where they are unaware of their surroundings. There is often a strange sense of quietness, even if they are being watched by thousands of spectators. There is a sense of spontaneity and ease, in which their reactions and movements are inevitably perfect. Without even expending much effort, they become capable of feats beyond their normal ability. A Formula 1 driver feels that they are flying along the circuit, effortlessly gliding past the other cars. A runner feels that they are floating along the track, easily beating their previous best time. A sportsperson reports that a ball (or a shuttlecock or a puck) appears much bigger than normal.

However, perhaps the most striking aspect of the Zone is that time seems to pass much more slowly than normal. In fact, this is the main reason why the Zone state brings peak performance – because an athlete has much more time to act and react. They have more time to respond to their opponents' movements or to their shots or tackles. They have more time to take evasive action or to avoid sudden unexpected events. They have more time to adjust or position themselves and to take their opponent by surprise. A tennis player may be on the receiving end of a 125-mph (200-kph) serve and yet perceive it moving so slowly that they can step back and hit an unbeatable return into the opposite corner of the court. A boxer may see their opponent's fist gliding so slowly that they're not only able to avoid the punch, but also to position themselves for an unexpected return blow.

Super-absorption

The Zone shares similarities with flow, when our attention is absorbed in stimulating tasks or activities. However, the Zone is a more dramatic (and rarer) state that alters our awareness in a more powerful way.

As we saw in Chapter 1, flow normally makes time pass faster, as well as bringing wellbeing and heightened energy. However, when absorption becomes especially intense, over a long period of sustained concentration, a more drastic change occurs. The state is no longer equivalent to flow; it becomes a different state of consciousness that I call "super-absorption". And super-absorption allows us to enter the Zone.

In some cases, an athlete builds up concentration gradually over the course of a game or contest. A racing driver or a golfer may concentrate hard for hours, eventually attaining a state of super-absorption and so entering the Zone. Here the game is akin to a meditation session, where a person gradually focuses their mind, attaining deeper states of stillness and wellbeing. In other cases, an athlete shifts quickly into super-absorption during a critical period of a game – for example, when they (or their team) are losing and making a concerted effort to catch up or in the final minutes of a game, when scores are tied or close. The intensity of the situation has a powerful focusing effect, rapidly intensifying concentration to the point of super-absorption.

It's as if, at a certain pitch of concentration, we pass through a portal into a different domain, like a ship passing through an estuary into the wide ocean. And in this different domain, along with major changes to our perceptions and our performance, we pass from the third to the fourth law of psychological time: time no longer passes quickly but stretches out drastically.

Athletes in the Zone

In every sport, there is a tiny proportion of elite athletes who are capable of feats that are way beyond even their most talented peers, some of which even seem to defy the normal laws of physics. What separates them from their peers is not simply their fitness, discipline, strength or tactics, but most of all, their regular access to the Zone. As the Finnish Formula 1 driver Mika Häkkinen said of his own experience of the Zone, "I'd say you have to reach this level if you want to be a champion."[1]

One of the greatest ever tennis players was the American Jimmy Connors, who was ranked World No 1 for 160 weeks consecutively in the 1970s and still holds the record for the most open championship titles (109). Connors' amazing success was due – at least in part – to his ability to enter the Zone. He described moments when the ball seemed to enlarge as it came over the net and to move so slowly that he had all the time in the world to choose his shots. Another example from tennis is Billie Jean King, who was the greatest female player of her time, winning Wimbledon six times, together with four US Open championships. To my knowledge, King has never directly described a TEE, but she has provided a beautiful summary of the Zone state. King wrote that when she was playing at her best, she became so focused that she was unaware of the crowd and even her opponent. All she was aware of was the ball and the face of her racket. "It's like I'm out there by myself," she said. "I appreciate what my opponent is doing, but in a detached, abstract way, like an observer in the next room . . . my concentration is so perfect, it almost seems as though I'm able to transport myself beyond the turmoil of the court to some place of total peace and calm."[2] It's highly likely that King experienced time expansion during these moments, too.

The American footballer John Brodie was equally familiar with TEEs. Playing for the San Francisco 49ers, Brodie was one of the leading American footballers of the 1960s and early 1970s, winning the NFL's "most valuable player" award in 1970. He described his experience of the Zone as follows:

> Time seems to slow down in an uncanny way, as if everything were moving in slow motion. It seems as if I have all the time in the world to watch the receivers run their patterns and yet I know the defensive line is coming at me just as fast as ever. I know perfectly well how hard and fast those guys are coming and yet the whole thing seems like a movie or dance in slow motion.[3]

TEEs can also occur during short duration events, such as sprints. The American sprinter Steve Williams – who equalled the men's 100- and 200-metre world records in the 1970s – described how, when he was running well, "10 seconds seems like 60. Time switches to slow motion."[4] Similarly, the short-length swimmer Courtney Allen described how, when she swam 50-metre freestyle races, "Every detail seems to last forever. When you replay it in your head, it just goes on forever."[5] It might seem that short-duration events don't offer much opportunity to generate super-absorption, but it's possible that athletes cultivate the state beforehand, through focusing and emptying their minds. The sheer intensity of the buildup to such events encourages a high pitch of concentration. While others might struggle to focus, the highest-level athletes may develop a mental readiness for super-absorption, which activates as soon as a race begins.

It's significant that many high-level athletes describe incredible powers of concentration, as noted by Billie Jean King above. Mika Häkkinen noted that for him, Zone experiences arise "with confident 100 per cent concentration and knowing exactly what you're doing".[6] In another example, the British golfer Tony Jacklin stated that he played his best games in "a cocoon of concentration" in which he was "living fully in the present, not moving out of it . . . I'm absolutely engaged, involved in what I'm doing at that particular moment".[7] This is precisely the intense and sustained concentration that generates super-absorption and allows athletes to enter the Zone.

Formula 1 in the Zone

As shown by the above examples, although any type of sport or game may give rise to TEEs, some sports are more closely associated with time expansion than others. One of these is motor racing. This is because the high speed and danger of motor racing, combined with the long duration of events (around 90 minutes in Formula 1 races) encourages intense, prolonged concentration.

The French Formula 1 driver Alain Prost vividly described a TEE during the 1986 Monaco Grand Prix. The Monaco circuit is notoriously difficult, with many narrow streets and sharp bends, but Prost felt as though "I was really flying and I could not see the speed. The speed meant nothing. It felt like I was driving at 30mph (48kph). I wouldn't describe it as a trance because that implies you're not in control . . . Your mind is still focused but it's really happiness."[8] Similarly, the Austrian driver Gerhard Berger recalled that when driving at his best, "Everything would be just like slow motion. Everything becomes very smooth and very soft, and you remember everything. When you're really on it, it's the best feeling in the world."[9]

In a more detailed example, a lower-level British driver called Mark Hughes – now a well-known motor racing journalist – described "one of the greatest days of my life" when he began a race in 26th position but finished third. Hughes achieved this remarkable feat by racing in an altered state of consciousness in which time disappeared. As sometimes occurs in falls, he experienced a strange sense of detachment, as if he was watching from outside his body. As he described it:

> It was something to do with the fact that I had nothing to lose and there was no pressure. But I remember thinking, "Wow, this is wonderful." Usually, you're so full on that you don't get to appreciate how fantastic it looks . . . I had the unmistakable sensation of being out of my body. It probably didn't last long but it felt like a long time. It's funny and it sounds weird but it felt unconnected to time . . . It's not really time . . . You felt you could go back, analyse and have a look . . . I felt like I could do anything and there was nothing stopping me from doing whatever I pleased. Afterwards I was walking on air and the feeling lasted for ages.[10]

The Scottish driver Jackie Stewart – who competed in Formula 1 during the 1960s and 1970s – has even suggested that time expansion is the essential prerequisite for success in racing. In a passage reminiscent of reports of preventative action during traffic accidents, he gives the example of a driver taking a corner at a 195mph (312kph) yet still performing complex manoeuvres:

> At 195mph, you should still have a very clear vision, almost in slow motion, of going through that corner –

so that you have time to brake, time to line the car up, time to recognize the amount of drift. And then you've hit the apex, given it a bit of a tweak, hit the exit and are out at 173mph.[11]

Some Examples from My Research

However, TEEs are certainly not confined to high-level athletes. In my collection, around 10 per cent of TEEs occurred in sports and games.

In a report that illustrates how super-absorption can arise during the last crucial moments of a game, a man described a TEE that occurred while playing ice hockey. In the final period, his team were down by a goal. His TEE began in the "face off", when the referee dropped the puck between him and an opponent:

In one gentle movement I pull the puck directly to the teammate behind me. I wasn't looking but I could hear the puck perfectly meet his stick . . . I move diagonally across the front of the net, keeping left because I can feel the shot about to come from behind. At this point everything has what would best be described as "tunnel sound". The shot rings off and, as I turn, my stick begins to raise itself into a shooting position.

As I'm turning, the rebounded puck reaches another teammate. He feigns to shoot and instead makes a smooth pass in my direction. My stick is still rising and I can see every dimple of the puck as it gently wobbles toward me. I watch it move just beyond the reach of some players and between the skates of others . . . The stick comes down in synchronistic timing with the path of the puck and the puck takes flight. The goalie dives toward my side of the net. The puck and the goalie seemed suspended in the air.

I seemed to be watching from a different position, not from behind the eyes, as I normally experience . . . The pipe sings as the puck smacks it on the way to the back of the net. The goalie collapses and there was a moment of pure silence. Followed by a volcanic eruption of celebration . . . The play which seemed to last for about ten minutes all occurred in the space of about eight seconds.

Similarly, a man described a game of table tennis that suddenly "turned into slow motion . . . I could see the ball and its flight and spin perfectly, anticipating its precise bounce, and position my body, arm, hands and wrist to hit perfect returns and sometimes from seemingly impossible positions where the ball was falling below table height and still hit a winner".

A skier described a "major moment of timelessness or at least time slowing down". This was also a state of effortless peak performance:

Skiing a fairly steep pitch doing short swing turns is something that requires quite a lot of skill and focus. Suddenly I was at one with skis, mountain, its slope, sky. . . . The motion was taking place totally effortlessly. Every flex of the knees, angulation of the ankles, planting of the poles, the skis carving the turns perfectly and adjusting to every undulation with perfect anticipation. I was a very competent recreational skier yet never had it felt this effortless.

The sense of oneness described by the skier illustrates the overlap between TEEs and spiritual experiences (which we will examine more closely in the next chapter). Similarly, an amateur motorcycle racer told me that she has had about "half a dozen" TEEs in races. She described how they have enabled her to "react and perform in what seems like a different dimension . . . I know I'm invincible and fully connected to Source. It's beautiful. [It] has lasted up to an hour and is the best spiritual experience of my life."

So far, I've described examples from physical or athletic sports, but TEEs can also arise from non-athletic games, such as computer games. Since they lend themselves strongly to flow, computer games usually have a time-contracting effect. A friend who spent a lot of his free time playing computer games told me that it was sometimes "scary" how quickly hours could pass. He told me that if he had time off work or wanted to savour his leisure time at weekends, he had now learned to avoid computer games in favour of activities that are less time-devouring. In this normal sense, computer games are a good example of the third law of psychological time ("Time seems to speed up in states of absorption").

However, occasionally the sustained concentration of computer gamers may intensify into super-absorption, opening a portal into time expansion. One person described how he had spent part of the evening watching an enthralling film then sat down to play a video game:

> For seemingly the next hour, I pretty much felt like some kind of prodigy player. Everything seemed to occur at about half the normal speed. I was able to look at other players actions, consider them and respond so rapidly that I found myself also able to consider many factors which usually I wouldn't have time to observe, let alone consider and react to. Single second events which usually feel fast felt slow.

Incidentally, I didn't receive any reports of sporting TEEs (or STEEs, as I will refer to them from now on) from the spectators of sports events. This surprised me at first. Anecdotally, I have heard some sports fans remark on how slowly time passes during intense moments of a game –

say the last few minutes of a cup final or during a penalty shoot-out. However, this effect probably doesn't reach the intensity of athletes' TEEs or of ETEEs, since watching a game involves a lower level of concentration than actually participating, and so is less likely to lead to super-absorption. This may not always be the case – it could well be that some people have fully fledged TEEs while watching nail-biting finales to games. (Let me know if you have!) Perhaps in most cases, though, there is just a mild time-expanding effect due to the inverse of the third law of psychological time ("Time seems to slow down in states of non-absorption"). For spectators, such intense periods of play are like the final stages of flights for nervous air passengers, inducing anxiety as well as a heightened awareness of time. In the same way that nervous flyers count down the minutes till touchdown, nervous spectators count down the seconds to the final whistle – and the seconds seem to pass slowly as a result.

ETEEs and STEEs

On the surface, STEEs may seem to have a different cause to ETEEs. In the last chapter, we noted that ETEEs are triggered by moments of dramatic shock or sudden unexpected events (like falls or car crashes), while in this chapter, we've seen that STEEs are caused by super-absorption. However, these two causes are really just two gateways into the same expansive timeworld, like two different entrances to the same building.

ETEEs and STEEs certainly share the same characteristics. Like ETEEs, STEEs feature a sense of calm wellbeing. As we saw above, Gerhard Berger described his STEE as "the best feeling in the world". Like ETEEs, STEEs often feature heightened awareness. This is partly why they confer an advantage, such as when a ball appears more distinct or larger or when a sportsperson clearly perceives the position and movements of other players. STEEs also feature rapid, detailed thought and action, enabling athletes to accomplish much more than would normally be possible. In ETEEs, this manifests itself in life-saving protective action; in STEEs, it produces extraordinary athletic feats. As with ETEEs, some STEEs include reports of external noises becoming "muffled" or silent, as when the ice hockey player above (see page 63) described "tunnel sound".

A further parallel is that some intense STEEs feature out-of-body experiences, like the most dramatic ETEEs. Many of the above reports mention a sense of detachment, as if the athletes were watching from above or in overview. For example, Billie Jean King related that, in her best moments, she felt like "an observer in the next room", while the ice hockey player alluded to this experience by saying, "I seemed to be watching from a different position, not from behind the eyes, as I normally experience." The racing driver Mika Häkkinen described this aspect in great detail. He reported that, when driving at his best, he felt like a "bird of prey" watching over his own car:

It's really like coming out of your body. You actually start to see more than just forwards. You learn to understand what's happening in the front, in the back and at the side. You can sense everything. It comes with confident 100 per cent concentration and knowing exactly what you're doing. There is nothing in life other than movement. Everything becomes like slow motion – even though you're going at unbelievable speed around the Monaco track. It's an amazing feeling.[12]

It's significant that Häkkinen associates this experience with "100 per cent concentration", emphasizing that super-absorption is the gateway to STEEs. His panoramic vision also exemplifies how the heighted awareness of the Zone benefits an athlete's performance.

In other words, ETEEs and STEEs are essentially the same experience, arising in different contexts. The important point is that super-absorption has the same effect as the shock of falls and car accidents: it induces an intense altered state of consciousness in which time slows down drastically.

A Permanent Zone?

I have suggested that the best athletes have regular, easy access to the Zone. But what if, beyond this, the Zone was actually their *normal* state while competing? What if an athlete was *always* in a different timeworld to their opponents, with all the advantages this confers?

This isn't just a hypothetical scenario. In my view, there have always been a tiny proportion of extraordinary athletes – the elite of the elite – who have experienced the Zone as their normal state. One example is the baseball player Ted Williams, whose career ran from 1939 to 1960. Williams is usually regarded as one of the greatest hitters (if not *the* best) ever, with the highest on-base percentage in baseball history (.482) – meaning that he reached base more frequently in games than any other player. (To put this in perspective, the average on-base percentage in major league baseball is around .300.) In fact, Williams is sometimes credited for coining the term "in the Zone", which he used to describe his own experiences. He claimed to be able to see the stitches on the seam of the ball as it flew toward him at 100 mph (160kph). Like Jimmy Connors, he also described how the ball appeared to grow, so that it seemed like a beach ball floating toward him in slow motion. Outside sport, Williams claimed he could read the labels on records spinning at 78rpm.

Scientists have dismissed Williams's claim of seeing the stitches of the ball as impossible. However, it only seems impossible in a normal state of consciousness, with a normal experience of time. Based on the many reports we have read over the last two chapters, it isn't at all implausible that Williams could accomplish this feat in an altered state, with an expansive sense of time. Certainly, this would help to explain his incredible prowess as a hitter. He is still the last hitter to average over .400 in a season, which he did in 1941.

The extraordinary Australian cricketer Don Bradman – who played international cricket from 1928 to 1948 – is a similar case. In statistical terms, Bradman is by the far the greatest batsman ever. A batting average of 50 runs per innings is usually seen as very high for an international cricketer, only attainable by elite players. However, Bradman's international career average was 99.94 runs per innings. (In his final innings, he required just 4 to gain an average of 100, but was out for 0, allegedly because he had tears in his eyes after a rapturous standing ovation from the crowd.) Only four other batsmen in history have ever averaged over 60 (the highest being 61.87). To use another statistical comparison, Bradman made 12 scores of 200 or over in his 52 international matches. The next best all-time figure is 11 times in 134 matches, by Kumar Sangakkara of Sri Lanka.

Because he stands so high above others in his field, some observers have referred to Bradman as the world's greatest ever sportsperson. One statistical analysis found that he far excelled extraordinary athletes in other fields, such as Pele, Jack Nicklaus and Michael Jordan. For example, the analysis found that Michael Jordan (many people's choice for the greatest basketball player ever, who no doubt routinely experienced the Zone too) would have had to average 43 points per game to equal Bradman, whereas he ended his career with an average of 30.[13]

The key to Bradman's incredible prowess has never been explained. He was physically unprepossessing, just 170 cms (5ft 7in) tall and of slight build. He never played cricket at a junior level and had no formal coaching. As a result, he had an unorthodox technique that would be frowned upon by modern coaches. He didn't play in a particularly attractive or flamboyant style either. For a large part of his career, he suffered health problems, including chronic muscle pain, later diagnosed as fibrositis (now referred to as fibromyalgia). An army medical during World War II found that he had poor eyesight, although he didn't wear glasses or contact lenses while playing cricket. Nevertheless, Bradman's contemporaries noted his incredible hand–eye coordination. He also placed a greater emphasis on physical fitness than other players of his time, which prepared him well for long stints at the batting crease. Perhaps most significantly though, Bradman was renowned for his incredible powers of concentration, which suggests that he may have routinely experienced super-absorption. Recently, it has even been suggested that he had autistic traits, which might help to explain his obsessive focus.[14]

Video footage of Bradman (of which there is sadly little) shows two striking aspects. First, there is the remarkable ease with which he plays, as if barely making any effort, even when he strikes the ball with great power. He seems to have a secret advantage, as if he's playing with a larger bat or ball. Second, there is the amount of *time* he seems to have to play his shots. Even when facing the fastest bowlers of his era, he always appears to have ample time to position himself and find the correct stroke. A cricket ball from a fast bowler takes around half a second to reach a batman, but it seems as if – like Ted Williams – Bradman was able to stretch this short period of time. As far as I know, Bradman never described an unusual state of consciousness or a slowed-down sense of time, but he's almost certainly an example of a sportsman who was permanently in the Zone.

This is also probably true of the footballer Lionel Messi, usually described as the best player of modern times. The Argentinian has won the Ballon D'Or – the global award for the footballer who performs best over a calendar year – a record seven times. He is renowned for his "impossible" goals that seem to defy the laws of physics. He threads through tiny spaces between groups of opponents, curls the ball more acutely than any other player, accelerates and decelerates with impossible rapidity, all while keeping the ball fixed to his feet at high speed. As the ex-England footballer Rio Ferdinand commented insightfully of Messi, "I just feel that in his own eyes and his own vision, the game just slows down for him. He plays in slow-motion because it comes to him so easy and so naturally."[15]

A small number of other footballers have had similar abilities, such as Messi's compatriot Diego Maradona and the Northern Irish and Manchester United player George Best. Such players seem to inhabit a different timeworld to their opponents. Gliding across the field, they miraculously keep their balance at high speeds, weaving around the clumsy lunging legs of defenders. Like people in ETEEs, they seem to have more time to position themselves and to respond to their opponents' movements.

To say that such players defy the laws of physics may sound like hyperbole, but in a sense it may be true. Classical physics states that time is fixed and absolute, flowing at exactly the same rate everywhere in the universe. However, more modern theories – such as Einstein's Theory of Relativity and quantum theory – suggest that time is flexible and indeterminate. Classical physics describes the world as it appears to our normal consciousness – a solid and orderly world which appears to be "out there", separate to us, which would exist in the same way even if we weren't here to perceive it. In intense altered states of consciousness such as the Zone, our perceived reality becomes more akin to the uncertain interconnected world described by quantum physics. So it may be when they operate in the Zone, athletes like Lionel Messi or Ted Williams do step outside the world of classical physics, apparently transcending normal concepts of velocity, space and time. (In Chapter 9, we will examine time from the perspective of modern physics in more detail.)

In view of the link between TEEs and altered states of consciousness, perhaps the key to understanding the extraordinary abilities of such sportspeople is that they *live* in – or at least always have easy access to – altered states. They don't need to cultivate super-absorption to enter the Zone, because they are always on the threshold of it. The Australian psychologist Michael Thalbourne coined the term "transliminal" to describe people with soft psychological boundaries, who are unusually open to altered states and anomalous experiences.[16] Such people are highly empathic, intuitive and creative, and have frequent psychic or spiritual experiences. In my view, this probably applies to extraordinary athletes like Don Bradman and Lionel Messi. Their transliminality means that they are always on the threshold of a different timeworld.

It's perhaps significant that some of the above athletes had or have unusual personalities. In some cases, they were troubled and unstable people. Ted Williams was a perfectionist with an uncontrollable temper, while George Best and Maradona had chronic problems with addiction. As noted above, Don Bradman – although less outwardly troubled – was possibly autistic, and his health issues were probably psychosomatic (at least to a significant degree). Lionel Messi doesn't appear to be a neurotypical person either. Some observers have suggested that he has Asperger's syndrome, due to his extreme shyness and his obsessive focus and concentration, although this has never been confirmed.

Transliminality is the key to the notion of the "suffering genius" – the artistic genius who is prone to depression and instability, like Van Gogh or Beethoven. Their soft ego-boundaries make them prone to altered states in which inspiration flows through them involuntarily, allowing them to rapidly create profound works of art, far superior to the more conscious and contrived work of their peers. At the same time, their soft boundaries may mean that they are highly sensitive, emotionally unstable and prone to depression. Sadly, this applies to some sporting geniuses too.

The *Siddhis*

Many factors contribute to sporting greatness – for example, physical fitness, strength, technical skill, tactical knowledge. But without entry to the Zone, such abilities only take an athlete so far. The real key to extraordinary sporting ability is the capacity to enter an altered state of consciousness, either through super-absorption or natural transliminality. And the most important feature of this altered state – in terms of contributing to a higher level of performance – is time expansion.

There is a connection here with what Indian Yoga philosophy refers to as the *siddhis,* extraordinary powers that develop through intense spiritual practice or which arise naturally as one undergoes spiritual development. Ancient Hindu texts describe a wide variety of abilities, including weightlessness, controlling material elements or natural forces, expanding or reducing the size of one's body, being undisturbed by hunger or thirst, and hearing and seeing distant events. Some of these abilities sound rather fanciful, it's true – perhaps there has been some exaggeration along the way, as is often the case with spiritual texts. However, there are many cases of spiritual adepts developing unusual mental abilities through intense meditation practice and other spiritual disciplines. One case was Swami Rama, a Yoga adept who, like many other Indian spiritual teachers of his time, moved to the US. In 1970, he was rigorously tested at the Menninger Clinic in Topeka, Kansas, where he demonstrated the abilities to dramatically change his heart rate, alter the temperature in parts of his body and generate different types of brain waves at will. He also demonstrated – under strictly controlled scientific conditions – the ability to move knitting needles by ten degrees with the power of mental intention.[17]

Perhaps we should view the abilities of exceptional athletes in a similar way – as superpowers that can be accessed through altered states of consciousness and cultivated through mental and spiritual training. Although they aren't mentioned in traditional yogic texts, abilities such as slowing down time or seeing objects as larger than normal could easily be viewed as *siddhis*. Whereas yogis cultivate altered states of consciousness through spiritual practice, extraordinary athletes do this through super-absorption, or because their transliminality means that they are always on the threshold of altered states. From this perspective, the philosopher Michael Murphy – a keen sportsman as well as student of Indian philosophy – has referred to sport as "Western yoga."[18]

As I suggested in the last chapter, sporting TEEs have an important bearing on the question of whether or not TEEs can be explained as a retrospective illusion. The fact that sportspeople gain such a massive advantage from their TEEs, with the ability to perform feats that would normally be impossible, shows that they must take place in the present.

But perhaps the most important implication of STEEs is the same as the one I identified at the end of the previous chapter: the fact the Zone is there, on the other side of our normal consciousness, is another indication that we human beings tend to underestimate ourselves. As with accidents, the Zone reminds us we are stronger and more capable than we assume, with vast ranges of potential that normally remain untapped.

CHAPTER 4
TIME OUT OF MIND
Awakening Experiences

After accidents and sport, there is one other major area where time expansion experiences occur: spirituality.

About 20 years ago, my wife and I went on holiday to Scotland. One grey and windy evening, I went for a walk along a coastal path. After a few minutes, I came to a bay and stopped to look at the scene. It was beautiful and dramatic, with the sea swelling wildly and giant waves crashing into rocks, spraying foam high into the air. I was completely alone except for a few seagulls flying over the beach. I kept staring at the cliffs, the rain-filled sky and the sea stretching endlessly into the distance.

After a few minutes, my awareness shifted. Everything around me seemed to come to life, as if an extra dimension of reality had been added. The sea was a living breathing sentient being, as it swelled and roared. The clouds possessed their own kind of consciousness, as they rolled by in the wind. Everything around me – the sky, the sea and the cliffs and even the rain and the air – was interconnected. There was a force or energy *underlying* them. They all stemmed from the same source, with a deeper unity beneath their superficial separateness. And I wasn't a detached observer of the scene, but part of the unity. I could feel my oneness with the sky and the sea and the rocks.

I sat down on a rock to stare at the sea. I was aware that I was observing a scene that had remained unchanged for eons and would remain unchanged for eons. I was here thousands of years in the past and thousands of years in the future. Time became spatial rather than linear, so that I was no longer confined to the present. The past and the future were part of the present, as if the present had stretched out to include them.

What Are Awakening Experiences?

The above could be described as a spiritual experience. However, I prefer to use the term "awakening experience", partly because the term "spiritual" is difficult to define and has different meanings to different people. For example, the term is often used with a strictly religious meaning, or in relation to paranormal phenomena like ghosts. (Another possible term is "higher state of consciousness", which I also sometimes use.)

I have spent a large part of my career as a psychologist studying awakening experiences. In fact, I originally became a psychologist because I wanted to study them from a psychological perspective. Based on my own experiences, I had a strong feeling that awakening experiences weren't essentially religious or necessarily connected to spiritual practices or traditions. They were natural human experiences that occurred in everyday situations to people who didn't consider themselves religious or even spiritual. I have led several research projects on awakening experiences and written extensively about them. (My earlier book *Waking from Sleep* was specifically about them).[1]

In awakening experiences – or spiritual experiences, if you prefer – normal human awareness expands and intensifies. Psychological processes that limit our normal awareness fall away, bringing a fuller and more intense vision of reality. We attain a state of perceptual clarity, similar to the bright and fresh awareness of young children (which we discussed earlier (see page 11) in relation to the laws of psychological time). As a result, our surroundings become more vivid and beautiful. Phenomena that we rarely notice become intricate and fascinating. However, perhaps the most striking feature of awakening experiences is a sense of *connection*. Objects and phenomena seem interconnected rather than separate. Our normal sense of separation dissolves too, so that we feel part of the world's interconnection. We feel intensely connected to other people (and living beings in general) with a strong sense of empathy, compassion or love.

Overall, awakening experiences bring a revelation that life is more meaningful and positive than we previously suspected. In comparison, normal human awareness seems limited, even delusory. Although the experiences are usually very brief (lasting no longer than a few seconds or minutes), they have a powerful transformational effect. People gain a new sense of trust and optimism, which sometimes leads to recovery from depression. For example, a woman who had an awakening experience after a period of intense psychological turmoil described "the most intense love and peace and knew that all was well".[2] In the aftermath of the experience, she found that the feeling of dread had disappeared from her stomach and felt able to cope again. "I looked around and thought about all the good things in my life and the future. I felt more positive and resilient."[3]

One of the main aims of my research has been to identify the different situations in which awakening experiences occur. The American psychologist Abraham Maslow believed that what he called "peak experiences" – which overlap with awakening experiences to some extent – are accidental and mysterious, while religious mystics usually believe that their experiences are given by the grace of God. But I have found that, while some awakening experiences occur spontaneously, in most cases they are linked to specific activities and situations. The most powerful triggers are psychological turmoil, meditation (or other types of spiritual practice) and contact with nature. Any activity that induces a state of deep relaxation or intense presence can lead to awakening experiences. This is why they are sometimes associated with listening to music and sex. They are also sometimes induced by psychedelic substances such as LSD or ayahuasca.

Awakening Experiences as Time Cessation Experiences

In relation to this book, the most salient aspect of awakening experiences is that they often bring significant changes to time perception. As powerful altered states of consciousness, they lead us into the expansive timeworld that lies beyond normal consciousness – the same timeworld we have already explored in relation to emergency and sporting TEEs.

At the same time, awakening experiences introduce us to some different features of this timeworld. While they often feature TEEs (particularly when induced by psychedelic substances), they also sometimes feature experiences of time *cessation*. Here, let me reintroduce the term "time cessation experience" or TCE.

TCEs will become more familiar to us over the next few chapters of this book. As we discuss them, we will see that they occur in two slightly different forms. In one form, time simply disappears, as if it expands to the point where it fades away completely. We step out of the flow of time, as if stepping out of a river. In the other variant, time doesn't strictly disappear, but becomes *panoramic* rather than linear. The past and the future seem to be laid out before us, alongside the present, as a landscape that we can survey. It's as if linear time slows down to the point where it simply stops, like a train that slowly pulls into a station. The scholar of mysticism Paul Marshall also identifies these two types of experiences, referring to the first as an experience of "no time", and the second is an experience of "all time".[4]

Mystical Time Transcendence

Awakening experiences have been reported all around the world, across different times and cultures. In the religious cultures of previous centuries, they were described as mystical experiences of union with God or an awareness of His divine power pervading the world, illuminating and uniting all things. Based on these experiences, many Christian and Islamic mystics claimed that God was not a personal being who overlooked the world, but a force or power that pervaded the world, and was also present within the human soul. As the 15th-century mystic St Catherine of Genoa wrote, "My being is God, not by participation only but by a true transformation and annihilation of my proper being . . . In God is my being, my me, my strength, my beatitude, my good and my delight."[5] Such views often incurred the wrath of religious authorities, who saw those who held them as dangerous heretics.

One example is the German mystic Meister Eckhart, who lived during the 13th and 14th centuries. He is one of the most remarkable mystics of all time, whose writings and sermons have an uncanny power which (appropriately) transcends time, so that they still seem lucid and resonant now. Meister Eckhart taught that the personal God was not the ultimate reality but emerged from an impersonal, formless source, which he called the Godhead. The human soul also derived from the Godhead, which meant that human beings are essentially one with God. As he put it, "the eye with which I see God is the same eye with which God sees me."[6] Not surprisingly, in 1326 he was accused of heresy. However, Meister Eckhart – by then in his late sixties – died before the Pope reached a decision on his case.

In passages that were surely based on his own experiences, Meister Eckhart described a state in which individuality fades away and we lose awareness of our physical form. In this state, time becomes panoramic, so that "there is only a present Now: the happenings of a thousand years ago and thousand years to come are there in the present."[7] In this time cessation experience, we view the world from the same perspective as God. According to Meister Eckhart, it was nonsense to "talk about the world as being made by God tomorrow, yesterday . . . Time gone a thousand years ago is now as present and as near to God as this very instant."[8] Time is one of the worldly qualities that separates human beings from God. As Meister Eckhart wrote, "Three things prevent a man from knowing God. The first is time, the second corporality and the third is multiplicity of number."[9]

Mystics have used phrases such as "the eternal now" or the "timeless moment" to describe TCEs. The 17th-century British mystic and poet Thomas Traherne departed from conventional Christianity by insisting that heaven was accessible to human beings here and now on Earth, rather than just in the afterlife. (The more easy-going attitude of the Anglican Church meant that he wasn't tried for heresy, although some of his contemporaries thought he was mad.) Traherne felt he had experienced heaven during childhood, when all things "abided eternally as they were in their proper places. Eternity was manifest in the Light of the Day, and something infinite behind everything appeared . . . The city seemed to stand in Eden, or to be built in Heaven." Traherne felt that he lost this awareness as he entered adulthood, due to the corrupting influence of the world. However, judging by his writings, he regained a heightened awareness in his later years, including a sense of transcending time. In his later works, he emphasized the presence of God in all things, including the human soul, creating a "Spiritual room of the mind [that] is transcendent to time and place". In this "room" time is not linear, but spatial. In Traherne's words, "all moments are infinitely exhibited" and "ages appear together, all occurrences stand up at once".[10]

A similar experience of time was described by a later British mystic, Richard Jeffries, who lived in the 19th century.

During his life, Jeffries was best known as a novelist and nature writer, but arguably his most impressive book is his "spiritual autobiography", *The Story of My Heart*. In this unusual book, he described an intensely awakened vision of the world, which seemed to be his normal state. As he wrote, "The fact of my own existence as I write, as I exist at this second, is so marvellous, so miracle-like . . . that I unhesitatingly conclude that I am always on the margin of life illimitable." Like Meister Eckhart and Thomas Traherne, Jeffries' mystical vision included a sense of timelessness. As he wrote, "I cannot understand time. It is eternity now . . . To the soul there is no past and no future; all is and will be ever, in now. For artificial purposes time is mutually agreed on, but really there is no such thing."[11]

In intense awakening experiences, time cessation is always accompanied with a transcendence of individuality and separateness. This is what Meister Eckhart means by saying that "corporality" and "multiplicity" – along with temporality – separate human beings from God. The illusion of time is linked to the illusion of separateness, and in mystical states we become aware of both the oneness of all things and the oneness of all time.

Examples from My Research

In my research, around a third of TEEs were from awakening experiences, which means that they are the most common form after accidents/emergencies. (Sporting TEEs are the third most common, but it made sense to discuss the latter straight after emergency TEEs because they are so closely interlinked.) Most of these "spiritual" TEEs, as we could call them, were linked to meditation or to a spontaneous state of presence. Around a third were linked to psychedelics. The psychedelic experiences were primarily extreme TEEs, while the meditative experiences were primarily TCEs.

Let's begin with some examples that involve an intense state of presence. Here a woman described an awakening experience at an outdoor pop concert. It was in a beautiful setting and she felt a sense of euphoria while dancing and singing along with the rest of the crowd. She also stepped outside the flow of time:

> This was enough to make us utterly present but halfway through the [concert] the sun began to set and created the most intense skyline awash with iridescent colours. It was so mesmerizing that the band stopped playing to comment on how magical it was. At that moment I felt a complete connection to everyone around me, the plants, trees and even the sky. The whole world, past, future, time itself slipped away and I was left forever in this moment of majesty.

Another woman described a powerful awakening experience that arose from an intense feeling of connection with a friend. She had talked with the man for two hours in a café and they were standing in the street, just about to go their separate ways:

> We are waiting for the green light to cross the street. And suddenly, everything slows down. Or more precisely, we slow down. It feels like we are in a different time-space, where there's no time. We are in a bubble, only the two of us, everything else, the buildings, traffic, other people and noise just disappear into a blurry, inseparable background. Everything in me is crystal clear, I feel all-consuming calm. The connection between us is heightened to the maximum degree and the experience is very intense. I know that we are part of the same experience, that we both understand it in the same way.

Similarly, a woman described an awakening experience that occurred during a calligraphy class: "There was music playing and I was focusing intently on the movement of my hand. I seemed to be transported to another place. I lost all sense of time and the sense of myself. There was a sense of my awareness blurring into a big expansive place that was peaceful." Another person described a TCE while listening to classical music, feeling relaxed and still, but attentive to the music. As she put it, "I heard the silence between the notes and it was breathtakingly beautiful."

Meditation

As I've already mentioned, most of the spiritual TCEs I collected were related to meditation. This requires a little discussion though, as there isn't a straightforward link between meditation and time expansion or cessation. In fact, probably the most common effect is that, *during* meditation – or at least during a good meditation, in which we keep our attention focused and don't have many distracting thoughts – time passes quickly. I go to a monthly group meditation session at my local Buddhist centre (although I don't class myself as a Buddhist, I enjoy the meditation and admire the Buddha's teachings). The meditation lasts 40 minutes and, almost without fail, I'm surprised when the bell rings to end the session. Sometimes it feels as if I'm just getting started. To me, those 40 minutes seem more like 20 or 25.

This effect has been confirmed by research. In 2022, researchers studied 22 long-term meditators, who reported that time passed faster when they were meditating, compared to when they were reading.[12] This fits well with the third law of psychological time ("Time passes quickly in states of absorption"). During a good meditation, very little sensory or cognitive information flows through our minds. Sitting quietly with our eyes closed, we process very little information from our surroundings. And with our minds fairly quiet, we process very little cognitive information.

Another variant is that *after* a good meditation, there is often a mild sense of time expansion. During a good meditation, our mental energy becomes concentrated and this additional energy takes over the reins of our normally automatic perception. Therefore, when we open our eyes after the meditation, our awareness is more intense, allowing us to perceive the suchness of the world. Here the second law of psychological time operates ("Time seems to go slowly when we're exposed to new environments and experiences."). Only here we haven't actually travelled to new environments, but have simply de-automatized our perception.

However, during an especially *deep* meditation, a very different experience may occur. In deep meditation, our self-system may dissolve away completely, bringing an experience of formless oneness. Our normal sense of identity disappears, and we transcend time. Here, for example, a man called Philip describes a TCE while meditating:

> [I had] no sense of a body, no sense of a Philip, no space or time, just profound awareness that knew it was simply awareness. A kind of remembering that this is what is and always had been and always will be ... There simply was no time and no experience in the normal sense of the word, which is time-based.

In another example, a man described a powerful meditative experience when he felt a sense of unity with the world: "I realized I was participating in that Awareness. Time stood still. The entire world seemed contained within me."

The shift from normal to deep meditation is similar to the shift that occurs in sports, when an athlete moves – usually via a state of super-absorption – from flow into the Zone. In both cases, we move from a swift passage of time (due to absorption) to a state of timelessness.

Are TEEs Awakening Experiences?

Incidentally, you might wonder if there is any essential difference between awakening experiences and TEEs generally. Aren't all TEEs – including the emergency and sporting TEEs we examined in chapters 2 and 3 – awakening experiences, to some extent?

Certainly, awakening experiences and TEEs share similar characteristics, such as wellbeing and heightened awareness. In addition, some awakening experiences feature the same strange quietness as some emergency and sporting TEEs, such as the woman who above described how "other people and noise just disappear into a blurry, inseparable background." However, there are significant differences too. Some essential characteristics of awakening experience aren't common in emergency and sporting TEEs, such as a sense of connection or oneness, and feelings of intense compassion. On the other hand, awakening experiences don't feature rapid, detailed thinking. In fact, they are usually associated with an absence of thinking.

These differences aren't surprising because of the different contexts in which the experiences occur. In most ETEEs or STEEs, there is a need for focused attention and rapid action in response to danger or challenge. In awakening experiences, attention is much more diffuse and there is no need to think.

However, in some cases, ETEEs are indistinguishable from awakening experiences, particularly when they involve a close brush with death. For example, you might remember the report in Chapter 2 of a man who was stung by a wasp (see page 40). In addition to stating that "Time was slow but also did not exist", he described "a feeling that [my friend], myself and the wasp were not separate. Nothing was." Similarly, a woman who witnessed an explosion described a powerful sense of oneness:

Extreme attention to detail, sound, movement. I was everywhere all at once . . . Aware of internal and external processes. As if I had a massive self, witnessing all. Watching people respond with incredible speed, strength and agility but in slow motion at the same time. Sound was loudly silent, if that makes sense.

These similarities aren't surprising either. ETEEs and awakening experiences are both intense altered states in which our normal psychological structures dissolve, allowing us to transcend normal awareness. Beyond normal awareness, however we arrive there, we explore the same wider landscape of reality. We experience different aspects of the landscape according to our situations and perspectives, but it's not surprising that the descriptions sometimes overlap.

In other words, ETEEs and STEEs *can* be equivalent to awakening experiences, but by no means always.

Psychedelic TEEs

One of the most effective – and direct – ways to transcend normal awareness and enter a wider landscape of reality is by ingesting psychedelics.

In 1937, the famous English novelist Aldous Huxley moved to California with his family and their friend Gerald Heard. Heard had begun to explore Eastern spiritual traditions and passed on his interest to Huxley. Huxley adopted the Hindu tradition of Vedanta – based on the teachings of the *Upanishads* and the *Bhagavad Gita* – and became a vegetarian and a regular meditator. In 1938, he formed a close friendship with the spiritual teacher, Jiddu Krishnamurti, who happened to be living nearby. Huxley's support helped Krishnamurti to become one of the most popular spiritual teachers of the second half of the 20th century.

Huxley's spiritual explorations inspired his 1945 book *The Perennial Philosophy*, a compendium of quotes and passages from Eastern and Western mystics, linked by his commentaries. However, Huxley never had a mystical experience himself until 1953, when he took mescaline for the first time. In a long essay about his mescaline experience, *The Doors of Perception*, he described "seeing what Adam had seen on the morning of his creation – the miracle, moment by moment, of naked existence".[13] Mundane objects like a chair and a pair of trousers seemed fascinating and beautiful, as if depicted by master painters. Certain that his vision was not a hallucination or distortion of normal awareness, Huxley stated, "This is how one ought to see, how things really are."[14] Like the mystics, Huxley also noted that psychedelics – the term that he coined, along with Humphrey Osmond – have a pronounced time-expanding effect. In *The Doors of Perception,* he noted that mescaline can "telescope aeons of blissful experience into one hour".[15]

The Doors of Perception had a massive influence on American counterculture, helping to inspire the hippie movement of the 1960s as well as a wave of self-exploration and self-development. (It also inspired the name of one of my favourite bands, The Doors.) In addition, Huxley's works inspired academic studies of the effects of psychedelics, although this initial research ground to a halt once LSD was made illegal. (Fortunately, research has resumed in recent years, since the easing of restrictions.)

Another British psychedelic pioneer was the author Richard Heron Ward. In 1957, he published a detailed account of his LSD experiences in his work, *A Drug Taker's Notes*. While Huxley noted an extreme expansion time under mescaline, Ward described how LSD induced an "absence of time". In a description reminiscent of Thomas Traherne's "something infinite behind everything appeared", Ward described an LSD experience in which he felt a "suggestion of the infinite as something eternally standing behind a life-time".[16] (In his introduction to the book, Ward also described a nitrous oxide experience that in one sense seemed to last "for an eternity" and in another sense "took no time at all".[17])

More recent research on a wide variety of other psychedelic substances, such as psilocybin, DMT, ayahuasca and salvia divinorum (a psychedelic Mexican plant), has confirmed these time-expanding and time-transcending effects. The Israeli philosopher Benny Shanon made a detailed study of the effects of ayahuasca, reporting the impression that "more time has passed than actually has. In extreme cases, it may seem that time has altogether stopped or that temporal distinctions are no longer relevant."[18] Elsewhere, Shanon describes a spatial perception of time under the influence of ayahuasca, in which the past and future exist alongside the present:

> Everything that has ever happened, as well as everything that will ever happen, all have an equal temporal status. In a certain sense, they are all there and one only has to look at them . . . A perspective is taken by which all that will have happened at all times is co-present. In this limit situation, the temporal may, in a fashion, be reduced to the spatial.[19]

The contemporary psychedelic researcher Peter Sjöstedt-Hughes has also noted that after DMT experiences, "one is often surprised and sceptical of the time suggested by the clock. Stepping outside of time is stepping outside of standard sentience . . . One enters what seems to be a state of excessive non-time, the eternal."[20] As the researchers Tim Bayne and Olivia Carter summarize, "One key feature of the psychedelic state is a distorted experience of time, with subjects typically reporting that time has stopped or slowed."[21]

Reports from My Research

Incidentally, we shouldn't assume from the above reports that psychedelics *always* induce awakening experiences. Certainly, not all – or even most – psychedelic experiences have spiritual qualities. Sometimes they just feature fascinating hallucinations or pleasant feelings of disorientation. More negatively, psychedelics may induce psychotic-like symptoms of paranoia and panic. But it's certainly true that, in the right context and with the right intention, psychedelics can induce many of the elements of awakening experiences, including time expansion.

In my collection of TEEs and TCEs, around 10 per cent are linked to psychedelics. One man reported how, during an LSD experience, he looked at the stopwatch on his phone and "the hundredths of a second were moving as slow as seconds normally move. It was really intense time dilation." He also reported a brief moment of timelessness, in which "time slowed down so much that I had no concept of time at all." Similarly, a man reported that he always found it hard to believe that his DMT experiences only lasted for a few minutes, since at the end "it's as if I've been somewhere for a very long time." A woman who took ayahuasca described how she felt that she entered a realm that was "endless and timeless, outside the normal space–time continuum".

There were also a few reports of TEEs caused by drugs that are not normally classed as psychedelics. For example, one man told me that he used to smoke a strong type of cannabis, that caused a radical slowing of time. As he described it, "The first few times I smoked [it], I had the experience that a thousand years passed just in a space of a few minutes. It was always a startling experience and one that seemed very real." Another participant told me that as a teenager he used to inhale butane gas, which "slowed time down and took me into sort of atomic realm. I'd say a few seconds was like hours."

All the above substances have powerful time-expanding effect for the same reason as accidents, the Zone state and more organic types of awakening experience (such as those induced by meditation or listening to music): they induce altered states of consciousness in which our normal psychological structures dissolve away, allowing us to enter a different timeworld. As we have seen throughout this book, as soon as we step outside our normal consciousness, we step into a different timeworld. In many psychedelic experiences, the change to our temporal experience is so extreme – usually bringing a massive expansion of time – simply because the consciousness-shift that psychedelics induce can be extreme.

There is probably a more minor factor, too. As we know that there is a link between information and processing and time, it's also significant that – in the poet William Blake's phrase, borrowed by Huxley – psychedelics open the "doors of perception". They remove the filter of familiarity from our minds, so that our surroundings become incredibly real and beautiful. Phenomena or objects that we don't normally pay attention to become objects of interest. As a result, our minds process much more information than usual, which no doubt contributes, if only on a less significant level, to their time-expanding effect.

The Temporal Effects of Other Drugs

As all drugs induce altered states of consciousness, does it mean that they all – including alcohol and opiates – have a time-expanding effect?

This is not the case. In fact, many types of drugs normally have the opposite effect, making time pass very quickly. In the 1960s, the American psychologist Stephens Newell conducted a wide-ranging study of the effects of different drugs on time perception. He found that many drugs, including heroin, tranquilizers and alcohol, had a time-contracting effect. Significantly, he found that the desire to speed up time – or to make it disappear – was part of the motivation for taking such drugs. In contrast, Newell noted that "the stronger psychedelics (LSD and others) have the effect of slowing or stopping time."[22]

Decades later, in 2012, the psychologist John Wearden – another of the UK's major time investigators – worked with Ruth Ogden and other researchers to investigate students' time perception under the influence of different drugs. Alcohol and cocaine were strongly linked to a swift passage of time. As one student reported, "when drinking on a night out . . . enjoying myself, time passed a lot quicker." Or as another noted, "When taking cocaine time always goes very, very quickly. It seems you start taking the drug and then, all of a sudden, several hours have passed and it is about 7am."[23] Obviously, the social aspect of these experiences is significant. Time tends to pass quickly in convivial social situations due to absorption. However, alcohol and cocaine seem to intensify this effect.

Wearden and his colleagues found more nuanced results with cannabis. When people smoked cannabis with friends – perhaps with alcohol or while watching TV – time seemed to pass quickly. On the other hand, if people smoked cannabis alone, time was more likely to pass slowly. As one student noted, "On cannabis, time goes slow, you think that an hour has passed when it was only 10 minutes." [24] Other studies have supported the latter findings – that under the influence of cannabis people tend to overestimate intervals of time, suggesting that time is moving more slowly.[25]

These discrepancies are due to the fact that, among all drugs, only psychedelics have an ego-dissolving and awareness-enhancing effect. In fact, this is precisely what distinguishes psychedelics from other drugs. Whereas other drugs create pleasant feelings by changing our brain chemistry, the primary effects of psychedelics are to intensify awareness and to dismantle our normal self-system or identity. There has been some debate over what category of drug cannabis belongs to – whether it is a psychedelic, a stimulant or a depressant – but perhaps the fact that it normally has a time-expanding effect adds to its claim to be a psychedelic.

In contrast, drugs like alcohol, cocaine or opiates leave our identity untouched. If anything, alcohol and cocaine seem to *strengthen* our self-system, reinforcing its boundary, leading to increased self-assertion and less empathy. This is probably the main reason why these drugs tend to make time pass faster, along with the social absorption (and therefore decreased information processing) that usually accompanies them. In addition, narcotics and sedatives (such as opiates) have a numbing, insulating effect. They reduce our awareness and mental activity, and weaken our connection to other human beings and to the world in general. As a result, our minds process significantly less information than normal, which contracts time.

So although, strictly speaking, such drugs induce altered states of consciousness, they don't weaken or dismantle our normal self-system and so don't bring time expansion. The altered states they induce are more superficial, like changing around the furniture in a house as opposed to moving (if only temporarily) into a different house.

Incidentally, one puzzling issue here is why psychedelics tend to induce extreme TEEs, while more organic and spontaneous awakening experiences – such as those induced by meditation, relaxation or music – are more closely associated with TCEs. I admit that I don't have a clear answer to this question (if you have any ideas, let me know!). However, it's perhaps significant that many psychedelic experiences don't involve a complete dissolution of the ego. Their primary feature may be a flood of sensory impressions or mental images, with perhaps only a partial dissolution of the normal self-system. More organic contexts such as deep meditation, which induce a state of mental quietness and emptiness, are probably more likely to lead to a state of complete ego-dissolution, and therefore to timelessness.

Ongoing Timelessness

Awakening experiences are temporary – they last for a few seconds, perhaps a few minutes, or at the most a few hours. But what if a person *lived* in a state of awakening, with a permanently heightened awareness and a continual sense of connection or oneness with their surroundings?

You might doubt whether this is possible. How could we concentrate on anything, while overwhelmed with feelings of oneness with the world and by the endless beauty of our surroundings? How would we do our jobs, pay our bills, drive the car, bring up our children and so on?

This definitely applies to psychedelics. If we were permanently dosed on LSD or ayahuasca, we would surely find it difficult to function as practical human beings. But the world's spiritual traditions have always insisted that it is possible to become permanently awakened (or enlightened) and at the same time live fruitfully. Almost every spiritual tradition features intricate paths and practices to cultivate an ongoing state of wakefulness, such as the eight-limbed path of Yoga, the eightfold path of Buddhism and the Christian mystical paths. These paths aim to cultivate wakefulness in an organic, integrated way, without the impairment and disorientation that psychedelics often bring.

In my research – as documented in books such as *The Leap* and *Extraordinary Awakenings* – I have found that wakefulness can arise outside the context of spiritual traditions. Just as with awakening experiences, it can occur in people who know nothing about spirituality – people who have never meditated, read spiritual books or heard of Buddhism or Daoism. Often, this occurs spontaneously following a period of intense psychological turmoil and trauma. In a phenomenon that I call "transformation through turmoil", a person's ego may dissolve away, either due to a sudden shock (such as bereavement or a diagnosis of cancer) or a long process of loss (perhaps due to addiction or depression).

In a minority of people, the breakdown of their ego allows a new self to emerge inside them – an awakened self with a more expansive and intense awareness, which becomes their new identity. Such people feel a powerful sense of inner wellbeing. The world seems more real and beautiful. They feel more connected to other people, with greater empathy and compassion. They also feel a strong sense of connection to nature. Their relationships become more authentic and intimate. They feel as if they have woken up, as if a veil has fallen away and they are living in a much fuller and more intense way.

Does this also mean that the temporal effects of awakening experiences – that is, an expansion or cessation of time – become normal? To an extent, this does seem to be the case. Here we can draw a parallel with the sporting TEEs we examined in the last chapter. We saw then that some high-level sportspeople can apparently slow down time at will, or even seem to experience a Zone state as their normal consciousness. In a similar way, some permanently awakened people may experience time expansion (or absence) as their normal state.

Sometimes people who undergo transformation through turmoil do have some initial difficulties adjusting to everyday life. This is especially the case if they don't fully understand what has happened to them, so that their feelings of wellbeing and connection are overlaid with a sense of confusion. They may feel overwhelmed by the new impressions and sensations flooding through their awareness. They may need to rest and retreat for a while, to allow their new psychological structures to stabilize.

They may also have difficulties related to time. Initially they may find it difficult to estimate the passing of time and to keep appointments and deadlines. For example, a woman called Donna – whose story I told in *Extraordinary Awakenings* – reported that after her transformation, "For about a year it was very hard because I was living in a different reality. Everything was very slow. Time didn't exist. Even when I was with my loved ones, I experienced them as a field of energy rather than as people."[26] However, in almost all cases that I am aware of (including Donna's), people do adjust to their new state, and begin to function well in the everyday world. In fact, they function at a much higher level than before – partly because they have a more harmonious relationship to time.

Some awakened people I interviewed commented on a sense that time seemed more expansive or seemed to be moving more slowly. As one Canadian woman – who underwent a transformation after a long period of stress and turmoil – told me, "When you're present all the time, every day seems full. A day seems to last for such a long time."[27] Others reported that they felt such an intense sense of presence that the concept of time seemed meaningless. As reported in *Extraordinary Awakenings*, a man called Parker underwent a shift after becoming so depressed that he seriously contemplated suicide, to the point where he stood on a bridge for three hours, trying to force himself to jump. After his awakening, he told me:

> It seems now as though the concept of time exists only in my conceptual mind. Time cannot exist in my direct experience of life itself – everything I perceive, everything I do, can only happen in the current moment in which it happens . . . This is fundamental to my current experience of no stress, worry, anxiety, sadness or depression – it's impossible to feel any of those experiences without slipping away from the present moment through thought . . . All my anxiety, stress and worry was based on this wandering mind. I don't do that at all anymore.[28]

Another research participant, Lynne, described a shift she experienced after losing her husband to cancer, followed by her mother's cancer diagnosis and death. She had a psychological breakdown, which eventually revealed itself as a spiritual awakening. In her words, she experienced "connectedness, intense love for everything and everybody, understanding, expanded awareness, curiosity and amazement at people and life generally."[29] In addition, she experienced the "eternal now" described by the mystics: "There was also an intense mind-boggling feeling of the past, the present and the future all existing at once, which again is very hard to explain . . . These feelings lasted very intensely and were constant for several years, and the shift of consciousness they engendered has been permanent."[30] Similarly, another person who had an awakening after a bereavement described how, immediately after her shift, her awareness was "outside of time, though some parts were translated into words or experiences in time . . . [T]he nature of time is that all time is simultaneous. This is how it is perceived from eternity."[31]

The key to being able to function in an ongoing state of wakefulness is that a person has the ability to focus their mind. When need be, they can focus their attention away from their bliss and oneness and concentrate on working, speaking, driving and other practical tasks. We could say that although they normally live outside time, they are able to slip back into time when necessary. In such moments, their wakefulness simply moves into the background. Lynne described this in relation to her own state, noting that "the business of everyday life pushes [my spiritual feelings] into the background and they come back more intensely when I'm alone or upset, or when I meditate."[32]

It's important to note that nearly all the experiences we have examined in this chapter feature a sense of oneness or loss of separate identity. As we heard from Meister Eckhart, there is a close link between transcending separation and transcending time. This suggests that our normal experience of linear time may stem from our normal sense of separation or duality. To a large extent, our normal sense of time seems to be created by our *self-boundary* – our feeling of being enclosed within our own minds and bodies, with the rest of the world out there, on the other side. When this boundary dissolves, time dissolves too. (We will discuss these ideas in more detail in Chapter 7.)

The above discussion also has an important bearing on a question we will consider toward the end of this book: is it possible to consciously change or control our perception of time? Or perhaps more specifically, can we permanently slow down our sense of time and live in an ongoing state of time expansion or even a state of time cessation? The cases of some sportspeople and some spiritually awakened people suggest that both these possibilities are viable.

However, before we turn to these other questions, we still have some ground to cover. We need to explore the nature of time in more detail, and the possibility that it may be illusory. And most imminently, we need to examine two further routes into an expansive timeworld: near-death experiences and the life review.

CHAPTER 5
DEEP TIME

Time in Near-Death Experiences

In 2009, Zak Khan was 39 years old and working as a college lecturer in England when he fell ill with severe symptoms. He was quickly diagnosed with leukaemia and admitted to hospital for treatment. One day, while a friend was visiting, he had a sudden seizure. His heart stopped beating for 2–3 minutes. During that time, while apparently unconscious and without any signs of life, he underwent a series of remarkable experiences that permanently transformed him.

When I spoke to Zak in 2021, it took him over two hours to describe his experiences. Even then, it was clear that he was skirting over some details, particularly those that were difficult to verbalize. As the seizure occurred, he "felt the most tsunami-like magnetic force pull on me inwardly. It was so giant in its strength that I knew it was futile to resist it . . . My breath shortened so much that I involuntarily gasped for air. I hear my friend panicking, shouting my name. As I hear her run out of the room, my consciousness is drawn inward."[1]

As his body shut down, Zak entered an otherworldly environment that seemed pervaded with harmony and peace. He was free of pain and felt a pure joy beyond any happiness he had known before. As he told me, describing the experience in the present tense:

> I feel like I'm lying on a very soft cloud in darkness. I can see shards of silver light in the darkness. I feel a sense of total peace, a peace beyond understanding that

completely embraces me. I feel totally comfortable. I begin to sink through the cloud. As I'm sinking, I start to wonder: what happens when I come out on the other side? Will I fall?

As I come through the cloud, I realize I'm on top of another cloud . . . I feel immensely free and realize that I have no physical form. I feel expansive and nebulous, with no pain and no restriction . . .

The next memory I have is of translucent, luminous light. It's incredibly pure. Immediately I know that this level of purity doesn't exist in the normal world. It's completely untainted, so pure, so full of loving-kindness that my whole being is weeping with gratitude. My whole body is imbued with translucent light. It's humbling beyond measure . . .[2]

Zak's visitor remained with him during the medical emergency and told him later that it was 2–3 minutes before the doctors resuscitated him. However, when I asked him how long he felt his subjective experience lasted, he told me, "It felt like at least seven or eight hours, perhaps longer. It's difficult to estimate because the concept of time seemed meaningless, completely irrelevant."

An American woman named Mary Neal had a similar experience. Neal is an orthopaedic spine surgeon, formerly the Director of Spine Surgery at the University of Southern California. In 1999, she miraculously survived a kayak accident in Chile, after being trapped under a waterfall for 12 minutes. Like Zak, while physically "dead", she left her body and watched herself from above:

I could feel my spirit peeling away from my body and my spirit went up toward the heavens. I was immediately greeted by a group of . . . somethings. I don't know what to call them. People? Spirits? Beings? I didn't recognize any of them, but they had been important in my life somehow, like a grandparent who died before I was born. They were so overjoyed to welcome me and greet me and love me . . . I had an overwhelming sense of being home.[3]

Neal felt the same sense of deep peace and harmony as Zak. With heightened senses, she perceived intense, radiant beauty all around her. She also reported an "absolute shift in time and dimension. I experienced all of eternity in every second and every second expanded into all of eternity."[4] As with many people who have such experiences, Neal felt so serene that she didn't want to leave, but the beings told her that she had to return, as she had work to do.

There was a tragic aspect to Neal's experience. The beings she encountered told her that her son would die before her, at a young age. Ten years later, this sadly came to pass, when her son was hit and killed by a car, aged 20.

The Puzzle of Near-death Experiences

Although extraordinary, Zak's and Mary's experiences are not uncommon. Such "near-death experiences" (or NDEs) have been intensively studied for over half a century.

In a sense, we have already looked at many examples of near-death experiences in this book. In Chapter 2, I provided reports of accidents and emergencies in which people had a close brush with death, such as falls and car crashes. However, in the present chapter the term "near-death experience" has a more specific meaning. It refers to when a person is neurologically and physiologically "dead" for a short period of time (or at least is biologically close to death, due to severely impaired physiological functioning, such as in a deep coma). Their heart stops beating, their brain registers no sign of activity and the other vital signs indicate death. However, on resuscitation, they report a remarkable series of conscious experiences.

Near-death experiences first came to wide public attention in the 1970s, with the research of psychologists such
as Raymond Moody, who published the influential book
Life After Life. Since then, due to the advances of modern
medicine (particularly in resuscitation techniques), reports
of NDEs have become widespread. A 2001 study by the Dutch
cardiologist Pim van Lommel, who began to investigate
NDEs after many of his patients talked about them following
resuscitation, found that 64 out of 344 cardiac patients
reported NDEs.[5] More generally, Professor Bruce Greyson of
the University of Virginia has found that between 10 and 20
per cent of patients who are close to death have NDEs.[6]

NDEs have several "core" features. Typically, they begin
with a feeling of separation from the body, sometimes with
a humming or a whistling sound. Then there is a journey
through darkness – sometimes through a passage or a tunnel
– toward translucent light. In the vast majority of cases, there
is a feeling of intense serenity, which is often linked to the
translucent light. As Zak reported, the light seems to have
a quality of overwhelming love or harmony. It seems to be
an all-pervading quality, even the source of all things. People
also report powerful feelings of connection, as if they are
no longer separate beings but one with the universe. These
aspects of NDEs overlap with high-intensity awakening
experiences, as described in the last chapter. These feelings
are so overwhelmingly positive that some people are
reluctant to return to their bodies and even feel disappointed
when they regain consciousness.

Other aspects of NDEs differentiate them from awakening
experiences. Like Mary Neal, people often describe meeting
deceased relatives or non-human beings, who they sometimes
describe as "beings of light". In a smaller proportion of cases,
people report life reviews, such as those which occur during
life-threatening falls or accidents. (We will investigate the
life review in more detail in the next chapter.) Throughout
the experience, people feel that their senses have become

heightened. In contrast to dreams or hallucinations, the experience feels much more real than ordinary experience.

NDE researchers have consistently found that they bring significant changes to time perception. As the British researchers Peter and Elizabeth Fenwick have noted, "Time is often changed in near-death experiences, and some people describe the period of the experience as being almost an eternity."[7] More specifically, according to Bruce Greyson's research, changes to time perception occur in three-quarters of NDEs.[8]

Fittingly, as they share many of the characteristics of intense awakening experiences, NDEs have essentially the same temporal effects as awakening experiences, albeit normally at a more intense level. NDEs can include both TEEs and TCEs, which sometimes overlap. In the case of TCEs, they may – again, as with awakening experiences – occur in two different forms. They may feature a sense of timelessness, in which a person feels they have stepped outside the flow of time. Or they may feature a "spatial" perception of time, in which the past and future seem to exist alongside the present, so that everything appears to be happening at once.

We'll look at these different aspects in turn, beginning with time expansion.

Time Expansion in NDEs

In Chapter 2, I suggested that ETEEs typically bring time expansion of the order of 10 to 40 (that is, one second is perceived as equivalent to 10 to 40 seconds of normal time). However, time expansion in near-death experiences seems to go far beyond this. For example, we've seen that, according to his friend, Zak's experience lasted 2–3 minutes of normal time. He experienced this period as 7–8 hours of subjective time, which is an expansion of the order of between 140 and 240.

In coma-related NDEs, a person may be close to death for days, with an extremely low level of brain activity. The same magnitude of time expansion may occur in these cases too, so that a few days become the equivalent of months or years. One example is the well-known case of the neurosurgeon Eben Alexander, who was in a coma for six days due to an acute bacterial infection in his brain. His doctors believed he had little chance of survival and that even if he did survive, he would suffer severe brain damage. Alexander's medical notes confirmed that he was extremely close to death and in a state of deep unconsciousness, with a bare minimum of brain activity. Despite this, he felt intensely awake and aware in a way that he had never been before, and underwent a series of life-changing conscious experiences. He recalled an "extraordinary odyssey that seemed to last for months or years – although it had all occurred within the seven days I lay unconscious in the hospital".[9] Although Alexander doesn't give a specific estimate, he seems to have experienced a similar degree of time expansion to Zak.

In some NDEs, time expansion may be even more extreme. In Chapter 7, which focuses on precognition, we'll look at the NDE of a woman called Elizabeth Krohn, who was struck by lightning. Her NDE lasted for around two minutes of normal time, but she felt that the extraordinary experiences she had while out of her body lasted for around two weeks.

Time Cessation in NDEs

Now let's turn to NDEs where time doesn't expand but disappears altogether.

An excellent example of a "timeless" NDE comes from a British man named David Ditchfield, who I interviewed for *Extraordinary Awakenings*. In 2006, he was seeing off a friend at a train station near Cambridge. He stepped on to the train to help his friend with her luggage and hug her goodbye, but as he stepped back off, his long coat got stuck in the train's closing doors. Unable to take off the coat, he found himself trapped as the train set off. Quickly losing his footing, he was pulled along the platform as the train gathered speed. Then he was pulled into the gap between the train and the platform and ended up on the train track, with the train hurtling above him.

One interesting aspect of this incident is that there was CCTV footage showing exactly how long it lasted. A total of 13 seconds passed between the moment the train started to move and when David was pulled down beneath it. But to David, "those 13 seconds stretched out like hours." In this sense, David had a classic emergency TEE of the kind we examined in Chapter 2. In fact, his report contains one of the clearest descriptions I have read of how time expansion offers the opportunity for life-saving planning and preventative action. As he told me:

I remember calmly making a plan and telling myself what to do. After all, it felt like I had all the time in the world to.

I recalled a news story about a baby boy who had been thrown from the third floor of a burning building. Incredibly, the baby survived the fall and the experts said it was because he was such a young infant, he was relaxed as he fell. So as the train began to accelerate and I could no longer run to keep up with it, I told myself to relax as my whole body was dragged into what looked like certain death, the gap between the train and the platform. And I did. And to this day, I believe the fact that I managed to relax my whole body as I was pulled under, was one of the reasons I survived this part of my accident . . .

I also remember making a plan as I was dragged down under the train, recalling adventure films like Indiana

Jones and James Bond, where the heroes always kept their heads down when trapped underneath speeding trains. So that's what I did. And again, I think that calm planning helped me to survive.[10]

As the final train carriage passed over him, David felt a surge of joy that he had survived, although now he became aware of intense pain. He noticed that the left sleeve of his coat had been ripped to shreds and then that his arm had been severed from the elbow downward.

This led to an even more powerful experience. Paramedics arrived quickly and rushed David to hospital. His life was hanging in the balance, as he was losing so much blood. While lying in the emergency department, he lost consciousness to the normal world, but became aware of an entirely different reality, similar to the environment described by Zak and Mary. There he felt embraced by a soft and warm darkness, with vivid colours and lights around him. There was no pain anymore and he felt very tranquil.

It was only a few minutes before David regained normal consciousness in hospital. He awoke to "an overdrive of noise and light and people and frantic voices" as he was rushed into the operating room. But in those few minutes he underwent a remarkable series of experiences that radically changed his personality and his perspective on life. As he told me:

In this new afterlife world, I saw an androgynous being standing at my feet and several dark-haired female beings on either side. Most wondrous of all, a waterfall of galaxies nearby and a tunnel of brilliant rotating flames of pure light appeared overhead, growing larger and larger as I stared into it . . .

In this new world, which seemed ultra-vivid and ultra-real compared to my day-to-day ordinary world, for the first time in my life, I experienced being part of an infinite perceptive field, a perceptive awareness that was outside of my usual stream of thoughts. I experienced myself being here, conscious of my surroundings, but at the same time my awareness felt massively expanded, a part of every single thing in my surroundings, as though my awareness was not just looking out through my eyes but was part of an infinite field of awareness. Now, I wasn't a separate point of perception, subjectively experiencing the surrounding field around me. I was this eternal field, and it was me. We were one and the same.[11]

When I asked David later how much time he felt had passed during the experience, he couldn't give an answer, since he felt he "stepped out of the flow of time completely. In this expanded existence, the notion of time simply didn't exist."

According to Bruce Greyson's research, this sense of stepping outside the flow time – or timelessness – is a feature of more than half of NDEs. Mary Neal reported this aspect, while also describing extreme time expansion. Another person who had an NDE after being electrocuted described an awareness that concepts of past, present and future were creations of the human mind, and how "at the earliest level of my NDE journey that human construct fell away, that arrow of time."[12]

Spatial Time in NDEs

One of the most well-known NDEs of recent years is the case of the author Anita Moorjani. In 2006, after suffering from cancer for four years, Moorjani was admitted into hospital for emergency treatment. She was told that it was too late to save her life, since the lymphoma had spread and metastasized throughout her body. Her vital organs shut down and she went into a coma, during which her NDE occurred. As with many of the NDEs I've described above, she found herself out of her body, in a state of a deep peace and clarity.

In NDEs, it's common for people to feel that they have a choice about whether to return to their bodies or to continue with the journey they have begun. As noted above, a part of them yearns to stay in this new realm of peace and beauty. However, they often feel a duty to return to their bodies and resume their previous lives, partly to share the wisdom they have gained. Moorjani felt reluctant to return to her weak, disease-ravaged body, but encountered her deceased father and her best friend (also deceased), who encouraged her to return and live her life fearlessly.[13]

In the days after she awoke from her coma, Moorjani's health miraculously recovered. For no apparent medical reason, her tumours shrank by almost three-quarters within four days. After five weeks, she was cancer-free. Although one of the doctors who treated her claimed that her recovery could have been due to the chemotherapy she received when she arrived at hospital, others doubted this was possible. One oncologist, Peter Ko, who visited Moorjani in November 2006 and examined her medical records, stated that chemotherapy could not have caused such a sudden recovery and would also have been highly dangerous, due to her failing organs.

However, from our perspective the most salient part of Moorjani's experience is that she had a TCE in which time became spatial rather than linear. She became aware that linear time is just a construct of the human mind and that in reality the past and future exist alongside the present. As she described her experience: "Time didn't run linearly the way we experience it here. It's as though our earthly minds convert what happens around us into a sequence; but in actuality, when we're not expressing through our bodies, everything occurs simultaneously, whether past, present or future."[14]

In a survey of NDEs by the American physician Jeffrey Long, just over a third of people reported specifically that "everything seemed to be happening at once."[15] In another example, Dr Gerard Landry, an anaesthetist at the University of Texas Health Center, had an NDE following a heart attack in 1979. He described finding himself in "a dimension beyond sequential time" where "past, present and future are all merged."[16]

Similarly, in his book *After*, Bruce Greyson relates the case of a 36-year-old policeman named Joe, who almost bled to death after surgery. During his NDE, he felt that he "knew what it was like to experience eternity, where there was no time . . . How do you describe a state of timelessness, where there's nothing progressing from one point to another, where it's all *there* and you're totally immersed in it?"[17]

As we saw in the last chapter, this spatial perception of time sometimes occurs in intense awakening experiences, as if linear time has collapsed and opened out, spreading in every direction rather than being concentrated in a single point. All events appear to be concurrently visible, rather than hidden away in the past or the future.

Personal Flashforwards

If time appears as spatial, with past and future events visible, then one might expect some NDEs to include reports of future events – that is, glimpses of personal or global events that are yet to occur. And indeed, although these are quite rare, they are a feature of some NDEs. I briefly referred to one at the beginning of this chapter, when Mary Neal was told that her son would die at a young age.

The NDE researcher Kenneth Ring refers to such experiences as "personal flashforwards". In some cases, people report being shown events from their future lives by beings or guides, while others report glimpses of their future life that appear like memories, passing through their awareness and lingering and returning to them after their NDEs. The events may not make sense at the time but are recognized when they occur in later life. Often, these visions are connected to the choice of whether to return from death – that is, people are aware that if they return to life, these are the events that will occur.

Ring describes the case of a man named Bill McDonald who had an NDE aged eight while severely ill in hospital, after his parents had been told that he was unlikely to survive. McDonald reported that "I was shown things and taken on a journey . . . I was shown things, and everything that I learned actually transpired 5, 10, 15, 20, 25, 30, even 50 years plus, later. In fact, I was shown a whole panorama of the major events in my life up to the age 59 . . . I'm seeing my future houses, jobs. I'm seeing my wife. I was eight years old. I hadn't met her yet! [I knew] I'd recognize her and I did. My wife, I saw my children . . . The next 50 years was like a constant déjà vu."[18]

In another case described by Ring, a woman had an NDE during childbirth, after her cervix was torn. She encountered beings who showed her visions of her future life. She saw a future scene in which she and her husband were middle-aged, living in a house she didn't recognize, with grown-up children who were visiting, and small children playing. This vision was extremely vivid – she could smell the salad she was making, her own perfume, soap from the shower, and freshly cut grass as her son mowed the lawn – and remained imprinted on her memory. When she described the experience to Long 22 years later, she told him that now she lived in the house she had seen, and that her husband and family – including her young grandchildren – were precisely as the vision showed.[19]

In a small number of cases, people claim to see visions not only of their personal futures, but of forthcoming world events. For example, Bill McDonald, whose NDE occurred in 1954, claimed to see visions of battlefields and helicopters from the Vietnam War, and also to witness the assassination of President Kennedy.[20]

There is so much about NDEs that contravenes the standard view of reality that your rational mind probably feels completely overloaded by now. Is it possible that these glimpses of the future were genuine? The problem is that these stories are impossible to verify. Our response to them depends on whether we accept the possibility of precognition or not – that is, the ability to sense or glimpse events from the future. If you don't accept this possibility, you'll probably conclude that the reports were due to embellishment, wishful thinking or coincidence.

I won't pass judgement on this topic now. I'll save this for Chapter 8, when we look at precognition in detail. (We will examine some other personal flashforwards in the next chapter, in the context of life reviews.)

Perhaps there is a simpler possibility: that NDEs are just hallucinations, caused by unusual brain activity when people are close to death. If this were the case, then we could also disregard these future visions as illusions too. This would also suggest that the temporal effects of NDEs are illusory, so it's an issue we should examine in detail.

Are NDEs Hallucinations?

We have seen that there is a correlation between the intensity of altered states of consciousness and the degree of time expansion. The most dramatic and dangerous emergencies tend to bring the most extreme time expansion. This also applies to the intense altered states induced by psychedelics, which – as we saw in the last chapter – also bring massive time expansion. In a sense, NDEs are altered states too – one of the most radical types that human beings can possibly experience – so it makes sense that they should bring such radical time expansion or time cessation.

In Chapter 2, I suggested that TEEs are not hallucinations, but real experiences that happen in the present. As we saw, this is clear from the fact that – as with David (see page 105) – people are able to think and act more rapidly than normal and to take preventative action that would normally be impossible in such periods. Far from being delusory, TEEs – like awakening experiences – reveal a more expansive reality, which appears to be *more* authentic than ordinary awareness.

But some altered states of consciousness *are* delusory, such as psychosis, drug-induced hallucinations, and many dreams. Perhaps this is true of NDEs too. They contain so many strange elements that it may seem logical to dismiss them as hallucinations, with no more significance than elaborate dreams.

However, it is extremely difficult to explain NDEs in conventional scientific terms. A variety of neurological and physiological theories have been put forward, all of which are highly problematic. I don't have space to assess all these theories here, so I will limit myself to two of the most popular.

Perhaps the most inexplicable aspect of NDEs is the idea that people can have conscious experiences at a time when they have no vital signs of life and no oxygen is flowing to their brain. Once a person's heart has stopped beating, the brain shuts down within 15–20 seconds. Since most mainstream scientists believe that conscious experience is produced by the brain, this doesn't make any sense.

The "Dying Brain Hypothesis" tries to solve this problem by suggesting that NDEs don't actually occur during the period when the person's brain is inactive but shortly before then, as a hallucination generated by a dying brain. Some of the characteristics of the NDE are – so the theory goes – linked to a lack of oxygen to the brain (cerebral anoxia), while the sense of wellbeing could be caused by the release of endorphins. A similar theory is that NDEs are hallucinations that occur as a person "comes to", either after a medical procedure as their anesthetic wears off or as they recover consciousness after a coma.

However, NDEs are completely different to the normal experiences that occur when the brain is starved of oxygen.

Cerebral anoxia brings chaotic, distressing and varied experiences, whereas NDEs are usually very clear and ordered experiences, featuring the same core characteristics (including, in most cases, an intense sense of wellbeing). The experience of "coming to" after an anesthetic or a coma is also completely unlike NDEs. In the former, mental functioning is usually very confused and impaired, lacking the clarity and alertness of NDEs.

Another popular theory is that NDEs occur when the brain releases natural psychedelic chemicals shortly before death. One candidate for this is DMT, whose time-expanding effects we examined briefly in the last chapter. DMT is unlike other psychedelics in that it is naturally produced in our bodies. However, normally our bodies only produce tiny amounts of DMT, nowhere near enough to produce changes in consciousness. However, there is a theory – put forward by the American psychiatrist Rick Strassman – that when a person is close to death, a large amount of DMT is suddenly released, creating the experience of the NDE.

DMT experiences are certainly more similar to NDEs than cerebral anoxia, often featuring the same spiritual aspects of heightened awareness, feelings of unity, altered time perception, a bright light and so on. However, DMT experiences do not feature any of the most pertinent content of NDEs, such as encountering deceased relatives, a life review and journeying through a tunnel. There is also no evidence that the brain releases a large amount of DMT when a person is close to death. (There are similar issues with another psychedelic candidate for NDEs, ketamine.)

There are also three more general points that argue against both theories (and to most attempts to explain NDEs in conventional terms). Firstly, there is the strong subjective sense of reality of NDEs. When we awaken from a dream or emerge from drug-induced hallucinations, there is usually a clear sense that we are returning to reality from a delusory state. We usually feel that the dream or hallucination was less authentic and reliable than normal consciousness. This also applies to altered states of consciousness that occur in cerebral anoxia or when we come to after an anesthetic. But in NDEs, the opposite is the case: they feel *more real* than normal consciousness. A review of over 1,122 NDEs by Jeffrey Long found that 95.6 per cent of people felt their experiences were "definitely real", with only three reporting that it was "probably not real" and only one saying it was "definitely not real".[21] As one person told Bruce Greyson, "Never, ever did I think it might have been a dream. I knew that it was true and real – more real than any other thing I've ever known."[22]

Secondly, there is the powerful transformational effect of NDEs. Once people have accepted and integrated their experiences, they almost always undergo a profound transformation. They usually become less materialistic and more altruistic, with a heightened sense of love and compassion for others. They feel more connected to nature. Their perception of their surroundings becomes more vivid, with a heightened sense of beauty. Some people develop new creative abilities. For example, David Ditchfield started to paint and to compose classical music after his NDE. Bruce Greyson has found that around a third of near-death experiencers change their careers, while three-quarters make significant changes to their lifestyle and hobbies. Three-quarters also said that they felt calmer and more willing to help others after their NDE, while two-thirds reported improved mood and higher self-esteem.[23]

However, studies of hallucinatory experiences caused by a lack of oxygen to the brain or temporal lobe stimulation don't show any significant after-effects. In fact, hallucinations in general are not transformational. They are usually quickly forgotten, skirting the surface of our minds without leaving any impact. It's true that psychedelic experiences sometimes have a transformational effect, but this is due to their spiritual content – the aspects that overlap with awakening experiences – rather than their hallucinatory aspects.

Finally, there is the matter of what NDE researchers call "veridical perceptions". It is common for near-death experiencers to report accurate details of what they witnessed while apparently "dead" or unconscious, usually during the early stages of the experience, while they were floating above their bodies. These include remarkably specific observations of medical procedures that were later verified by the medical professionals. In a review of 93 reports of out-of-body experience during NDEs, Bruce Greyson and Jan Holden found that 92 per cent were verified as accurate by external sources.[24] (A recent study of NDEs called *The Self Does Not Die* reviews more than 100 cases. You can also refer to my book *Spiritual Science* for summaries of such cases.)

As one final point, it's significant that physicians are often convinced of the authenticity of the NDEs that their patients report. In fact, one of the striking things about NDE researchers is that they are usually not parapsychologists, but medical professionals. The British researcher Penny Sartori worked as an intensive care nurse for 17 years and began to research NDEs after many of her patients reported them. Pim van Lommel spent 26 years as a cardiologist. Jeffrey Long is a radiation oncologist. Similarly, one of the most renowned American NDE researchers, Dr Martin Sabom, was a cardiologist who specialized in resuscitation.

If NDEs are real experiences, they have many important implications. Perhaps the most important is that human consciousness can continue in the absence of brain activity, which in turn implies that consciousness is more than simply a neurological process. Even more controversially, NDEs also suggest the possibility of life after death. Admittedly, a continuation of consciousness during a short period of brain death doesn't necessarily imply *permanent* survival of our individual consciousness. However, it's significant that almost everyone who has an NDE becomes convinced of an afterlife. They believe that they have briefly experienced death and lose their fear of it. According to Bruce Grayson's research, 86 per cent of participants reported less fear of death.[25]

However, in relation to this book, the important point is that if NDEs are not hallucinations, then their temporal aspects are not illusory. Rather than the result of aberrational brain activity, the time expansion and cessation of NDEs may be a glimpse of the fundamental (or at least *more* fundamental) reality beyond our normal narrow human awareness. It may be our normal fast-flowing linear experience of time that is aberrational and illusory. The true nature of time may be expansive and spatial, rather than constricted and linear.

Time and the Process of Dying

Of course, we will all have a near-death experience at some point, even if it doesn't happen until old age. And evidence suggests that many of us will experience time expansion and/ or time cessation when we undergo the process of dying.

In 2015, the Swiss psychotherapist Monika Renz led a team of researchers who investigated the experiences of 680 dying people in hospices. The research found that, as they were dying, many people transitioned to a higher state of consciousness, moving beyond anxiety and pain and into acceptance and peace. Based on the findings, Renz has identified three stages of the dying process: pretransition, transition and post-transition. In the post-transition phase, the dying person lets go of fear, accepts death, and experiences a state of ego transcendence and tranquillity. In this stage, they experience the same sense of bliss and oneness that occurs in NDEs and intense awakening experiences. In Renz's words, "Patients feel free and at the same time somehow connected, a connectedness with the universe, with a transcendental sphere."[26]

The post-transition phase also features time expansion and cessation. As Renz has described it, during this stage "there is an utterly different atmosphere, a state beyond time, space and body . . . This makes such moments seem eternal."[27] Renz found that more than half of the dying people proceeded to the post-transition phase and so experienced a spiritual opening. She believes that the true figure may have been higher but that a number of other patients were "either unable or too shy or too tired to contribute" to her study.

Even more importantly, though, Renz's findings suggest that for most of us, death will be a peaceful and even blissful process. One theme that has emerged from this book is that many situations that we dread turn out to be less distressing than expected. As we saw in Chapter 2, although we naturally fear accidents and emergencies, we often experience them as powerful positive experiences that uncover depths of resilience and confidence we never realize we possessed. In that chapter, we saw that this even applies to life-threatening situations such as falls, when people sometimes feel a surprising peace and joy.

The material we have examined in this chapter reinforces this paradox. NDEs generally, together with Renz's research, suggest that the human fear of death – most people's greatest fear – may be misplaced.

CHAPTER 6
THE LIFE REVIEW

When Life Flashes Before Our Eyes

In 1791, a 17-year-old Irish sailor named Francis Beaufort – later Rear Admiral Sir Francis Beaufort, creator of the Beaufort Wind Scale – fell into the sea while trying to fasten a small boat to a ship. Unable to swim, he quickly started to sink. Many years later, Beaufort wrote a detailed description of the experience that ensued, as he came close to death through drowning.

Beaufort first described "a calm feeling of the most perfect tranquillity . . . I no longer thought of being rescued, nor was I in any bodily pain." He compared his state of mind to the "dull but contented sort of feeling which precedes the sleep produced by fatigue". However, while his senses felt dulled, his mind became incredibly alert and active, "invigorated in a ratio which defies all description; for thought rose after thought with a rapidity of succession that is not only indescribable, but probably inconceivable by anyone who has not been himself in a similar situation". A wide range of topics sped through Beaufort's mind, including how his father and other relatives would respond to his death and a "thousand other circumstances minutely associated with home".

However, Beaufort's account is most significant because he provides what is probably the first ever detailed account of a "life review". His whole past life opened up before him, as if in a panorama, beginning with the most recent events:

The course of those thoughts I can even now in great measure retrace . . . our last cruise – a former voyage and shipwreck – my school, the progress I had made there, the time I had misspent and even all my boyish pursuits and adventures. Thus, traveling backward, every incident of my past life seemed to me to glance across my recollection in retrograde procession; not, however, in mere outline as here stated, but the picture filled up with every minute and collateral feature; in short, the whole period of my existence seemed to be placed before me in a kind of panoramic view, and each act of it seemed to be accompanied by a consciousness of right or wrong or by some reflection on its cause or consequence – indeed many trifling events, which had been long forgotten, then crowded into my imagination, and with the character of recent familiarity.

Beaufort was unsure how to interpret his experience, except that it indicated that "death is only a change or modification or our existence, in which there is no real pause or interruption." He was amazed that so many thoughts and experiences could flood through his mind in the short period before he was rescued by fellow sailors. In his estimate, "certainly, two minutes could not have elapsed from the moment of suffocation to the time of my being hauled up."[1]

Types of Life Review

We have already looked at examples of the life review in Chapter 2. We saw that, in 1871, Albert Heim had a very similar experience to Beaufort, when he slid off a cliff. While experiencing the same calm contentment as Beaufort, he "saw my whole past life take place in many images, as though on a stage at some distance from me".[2] We also heard the story of a man in a car crash who "recalled every experience I'd ever had at once, all memories and feelings and emotions. It included all kinds of things that I'd completely forgotten about."

Life reviews are, of course, the origin of the phrase "my life flashed before my eyes." At the beginning of the last chapter, I mentioned that the term "near-death experience" can be used in two different ways: to describe any dramatic close encounter with death such as a fall or an accident, or to describe the specific type of experience that may occur when a person "dies" for a short time but continues to be conscious (which is the meaning I adopted in the last chapter). The life review can occur in both types of near-death experience. In terms of the first type of NDE, the life review is most common in falls and near-drowning. In the second type of NDE, the life review occurs alongside other aspects such as an out-of-body experience, a journey toward light, meeting strange beings or deceased relatives and so on.

To my knowledge, there is no reliable estimate of how frequently life reviews occur in life-threatening situations (although the fact that they have given rise to a common phrase suggests that they are common). However, we have reliable data showing that around 20–25 per cent of the *second type* of NDEs feature a life review.[3]

The life review occurs in three slightly different forms. Around a half of life reviews feature a complete sequential vision of all the events of a person's life, which is often compared to a high-speed film (or a play, in Heim's version). In this type of life review, it's important to note that when people claim to relive *every* event of their life, they're not using rhetorical exaggeration (as when a person might say, "I've travelled all over the world.") People literally *do* mean this. For example, an Australian woman called Gill Hicks was severely injured in a terrorist attack in London in 2005. While floating through the air, looking down on her lifeless body, "My life was flashing before my eyes, flickering through every scene, every happy and sad moment, everything I have ever done, said, experienced. It was all being played like a film running at high speed in my head."[4] Usually, the sequence moves forward, from childhood to adulthood and onward, but sometimes it moves in reverse, from the present moment back through the past. As Hicks' wording suggests, it's this type of life review that gives rise to the phrase "my life flashed before my eyes."

In the second variant of the life review, about a quarter occur as a panorama rather than a sequence, as in Francis Beaumont's example. In this variant, strictly speaking, life does not "flash before our eyes". The events of our lives are simply *there*, spread out before us. Our attention roams over them as a climber surveys a landscape from the top of a mountain. Here is a good example from one of the patients of the Dutch cardiologist, Pim van Lommel. The life review occurred during an NDE following cardiac arrest:

> All of my life up till the present seemed to be placed before me in a kind of panoramic, three-dimensional review, and each event seemed to be accompanied by a consciousness of good or evil or with an insight into cause or effect . . . Looking back, I cannot say how long this life review and life insight lasted; it may have been long, for every subject came up, but at the same time it seemed just a fraction of a second, because I perceived it all at the same moment. Time and distance seemed not to exist. I was in all places at the same time, and sometimes my attention was drawn to something and then I would be present there.[5]

The remaining quarter of life reviews feature just a *selection* of a person's life events, rather than all of them. This usually consists of events that hold special significance or had a powerful emotional impact. In one example from my collection, a participant recalled a violent attack when she was beaten unconscious: "I watched screenshots or snapshots of some of the best moments of my life. It was like a film playing with the soundtrack of people screaming and shouting from the violence." Similarly, for my book *Out of the Darkness*, I interviewed a man called William Murtha who almost drowned when a freak wave swept him into the sea. While drifting in and out of consciousness in the water, William "was given a replay of almost ten to a dozen past experiences, which I now have defined as probably the most emotionally traumatic times of my life. It was almost like I was being replayed them to make me see them from a higher, alternative perspective."[6]

There is an excellent example of this type of life review at the end of the film *American Beauty*. After being shot dead by his neighbour, the film's main character, Lester Burnham, finds himself floating above his house and his town, serenely reliving some of the pivotal moments of his life:

> I had always heard your entire life flashes in front of your eyes the second before you die. First of all, that one second isn't a second at all, it stretches on forever, like an ocean of time . . . For me, it was lying on my back at Boy Scout camp, watching falling stars. And yellow leaves from the maple trees that lined our street. Or my grandmother's hands, and the way her skin seemed like paper. And the first time I saw my cousin Tony's brand-new Firebird . . .

Salient Aspects of the Life Review

It's important to note that the life review is not just a replay of memories. People don't just remember events but *relive* them. They don't just recall them from the vantage point of the present, they experience them *in* the present, as if the events are still happening.

People often report that the life review is much more vivid and detailed than ordinary memory. Rather than simply recollecting events, they relive every thought and conversation connected to the events. Because their awareness is more intense, they often describe details that they were unaware of at the time of the original experience. For example, a man named Tom Sawyer had an NDE while working underneath a pickup truck, which collapsed on his chest. During a life review, he relived the experience of mowing the family lawn for the first time, aged eight. He described how he "relived every exact thought and attitude; even the air temperature and things that I couldn't have possibly measured when I was eight years old. For example, I wasn't aware of how many mosquitoes were in the area. In the life review, I could have counted the mosquitoes. Everything was more accurate than could possibly be perceived in the reality of the original event."[7]

In the life review, awareness also becomes much more *wide-ranging* than normal. People witness events in overview, like birds flying high above a landscape, rather than looking from a single vantage point on the ground. As the quote from Pim van Lommel's patient above suggests, space and distance seem to disappear, as well as time. There is an acute awareness of the consequences of one's actions, with their wide-ranging rippling effect.

For example, a woman who had a life review during an NDE caused by a miscarriage described it as:

[A] total reliving of every thought I had ever thought, every word I had ever spoken and every deed I had ever done; plus the effect of each thought, word and deed on everyone and anyone who had ever come within my environment or sphere of influence whether I knew them or not (including unknown passers-by on the street); plus the effect of each thought, word and deed on weather, plants, animals, soil, trees, water and air.[8]

Even more striking, some people report a telepathic awareness of other people's thoughts and feelings in response to their actions. Tom Sawyer described how, as a mischievous eight-year-old, he cut down a patch of weeds even though he knew his aunt (who lived at the back of his house) wanted to let some wild flowers grow from them. As he told Bruce Greyson:

> I also experienced it exactly as though I was Aunt Gay, several days later after the weeds had been cut, when she walked out the back door. I knew the series of thoughts that bounced back and forth in her mind . . . between thinking of the possibility [that I had done this], and saying to herself, "Well, it is possible. No, Tommy isn't like that. It doesn't matter anyway, I love him. I'll never mention it" . . . Thought-pattern after thought-pattern. What I'm telling you is, I was in my Aunt Gay's body, I was in her eyes, I was in her emotions, I was in her unanswered questions.[9]

This awareness of the effect of one's actions is a particularly salient aspect of the life review. In cases where people have behaved poorly and hurt others, there is often a sense of guilt or embarrassment, bringing a moral awareness of the importance of treating others with respect and kindness. As Francis Beaufort put it, each recollected event or act is accompanied with "a consciousness of right or wrong, or by some reflection on its cause or consequence". As one woman who had an NDE after suffering respiratory complications told Bruce Greyson, "I could feel everything everyone else felt as a consequence of my actions. Some of it felt good and some of it felt awful."[10] Another person reported that although the life review was a generally a pleasant experience, she felt "so very sorry for certain things I had said or done. I hadn't just seen what I had done, but I felt and knew the repercussions of my actions. I felt the injury or pain of those who suffered because of my selfish or inappropriate behaviour."[11]

Time in the Life Review

As with NDEs, there are so many dramatic aspects of life reviews that it's easy to forget their significance in relation to time. However, they feature a level of time expansion far beyond any we have examined so far. How is it possible to relive an entire lifetime in a few seconds? How can decades of detailed experience, including the wide-ranging consequences of our actions and the thoughts and feelings of others, become so compressed?

In sequential life reviews, people often comment on the extraordinary speed at which images flash by. One woman who had a life review after respiratory complications told Bruce Greyson, "The information was following at an incredibly breakneck speed that probably would have burned me up if it weren't for the extraordinary Energy holding me."[12] A man who had a life review during an NDE at the age of 12 calculated that his life up to that point had lasted around 378 million seconds, which "were compressed to just a few seconds in the life review, which included moral lessons. That's like superluminal light speed. I am aware it sounds impossible."[13]

As the above calculation shows, time expansion in the life review is even more extreme than NDEs or in the most intense psychedelic experiences. We've estimated that in ETEEs, time expands to the order of 10 to 40. We've seen that in NDEs, time may expand much more radically, so that two minutes passes like several hours, or even several days. However, to experience 12 years in "just a few seconds" is an expansion of the order of tens of millions. Whereas in a typical ETEE, a few seconds may stretch into one or two minutes, in this life review a few seconds expanded to 12 years. Or to put it another way, each second became longer than a year.

Panoramic life reviews are easier to comprehend in temporal terms. There is no need to explain how years of time can be compressed into seconds, because time has no speed at all. As in the "eternal now" of mystical experiences or some NDEs, we simply step out of the flow of time. Time becomes spatial rather than linear, and we view all the events of our lives with the same type of instantaneous sweeping vision as faces in a crowd or the features of a landscape.

In panoramic life reviews, people often comment on the spatial nature of time, or the fact that it doesn't seem to exist at all. Gigi Strehler, founder of the Near Death Experience UK community (which offers support to people who have had NDEs), experienced a panoramic life review in 2011 after a medical emergency. As her physical pain faded and her body shut down, she found herself in a timeless realm. As Gigi told me:

> The fabric of time just curled and folded back into itself. I had no defining existing point in relation to time. There was no beginning or ending. I was in a place where time isn't calculated. I could suddenly see everything I'd ever done in my life, good and bad. I experienced everyone's interactions with me as a unity, every single energetic transmission.
>
> When I was resuscitated, I felt I had returned from eternity. It was the peace that passeth all understanding. Even now, here, as I'm talking about it, I'm not able to fully latch on to the content.

As with many people in NDEs, Gigi became aware that time is a creation of the human mind: "Time is not linear, in fact it is a man-made concept . . . [In NDEs] you experience all that was, is and will be in one simultaneous moment that you are a part of and at one with."[14]

The Jazz Musician Who Glimpsed His Future Life

There is another significant temporal aspect of life reviews: the fact that they sometimes incorporate life *previews*. In Kenneth Ring's term – as discussed in the last chapter –they may include personal flashforwards.

Tony Kofi is a leading British jazz musician and saxophonist, twice winner of the BBC Jazz Awards. Before he took up music, Kofi was an apprentice builder. One day, aged 16, he fell from the third storey of a building. "Everything became very, very slow," he recalled in a BBC interview, describing the time expansion that is typical of falls. As is also typical, he saw a cascade of visions. But rather than recognizing them as past events, he sensed that he was being shown visions of his future. As he described it, "In my mind's eye I saw many, many things; children that I hadn't even had yet, friends that I had never seen but are now my friends. The thing that really stuck in my mind was playing an instrument." Tony also saw images of different places around the world, although at the time he had never been out of the UK.

Tony landed on his head and lost consciousness. When he came to in hospital, he felt like a different person and didn't want to return to his previous life. Although he loved listening to music, he hadn't had the option to study music at school and had never played an instrument. No one in his family was musical either. Over the following weeks, the images kept returning to his mind. As he described it, "Every time I closed my eyes, the images were there." Most of all, the image of him playing an instrument kept returning. He felt that he was "being shown something", and that the images represented his future.[15]

A few weeks later, Tony saw a picture of a saxophone and recognized it as the instrument from his visions. On receiving some compensation money for the accident, he bought a saxophone. He practised for hours a day, teaching himself by playing along with records. His parents were confused, as he hadn't told them about his vision. They tried to discourage him, disappointed that he was giving up his apprenticeship for the pipedream of becoming a musician. He tried to gain a place at a UK music college, but as he had no formal music qualifications, no one would accept him. However, he was so determined to become a musician that he applied to American colleges and was accepted as a self-taught musician by the Berklee College of Music.

Over the course of his career, Tony has travelled to all the places he saw during his fall. He now has three children, whom he believes he saw during his fall.

Exploring the Life Preview

Is it really possible that Tony Kofi glimpsed his future? Did he see himself playing the saxophone because his future as a musician was somehow already determined? Or did his vision simply encourage him to take up the saxophone, in a case of self-fulfilling prophecy? Similarly, is it possible that the children he saw were his own future children, and that he saw his future friends and places he would later visit? Or did he simply convince himself of this many years later? After all, surely his memory of faces and places would have faded after so long?

Tony Kofi's story is not unusual, though. According to Bruce Greyson's research, around a third of life reviews feature future visions As with the personal flashforwards in some NDEs, they may seem to make little sense at the time. However, people often report – like Tony Kofi and again, as with flashforwards in NDEs – that the events come to pass eventually. In a case from Pim van Lommel's research, a person described a life review in which:

> In a flash I saw the rest of my life. I could see a large part of my future life; taking care of my children; my wife's

illness; everything that would happen to me, both in and out of the workplace. I could see it all. I foresaw my wife's death and my mother's passing. One day I wrote down all the things I saw back then: over the years I've been able to tick them all off. For instance, I saw my wife on her deathbed, wrapped in a white shawl, just like the one she was given by a friend of hers shortly before she died.[16]

The idea of glimpsing the future may seem incredible but, as I pointed out in the last chapter, it follows logically from the non-linear view of time we have discussed. If linear time is an illusion and if time is actually spatial, then perhaps the future already exists in some sense. And if, in a spatial vision of time, the whole of a person's past life is laid out before them in a panorama, the future may also appear in a panorama.

Of course, the idea that the future may already exist has implications in terms of free will. Does it mean that our lives are predetermined? Are we powerless to change the future, even if we feel as if we are using free will to make choices?

This is a topic we will explore in more detail in the final chapter of this book, but for now it's worth noting that in some life previews people feel they have a degree of choice over their future. While some events are presented as unavoidable – as definite as the present – others are viewed as *potential* events that can be avoided or realized, depending on their choices. For example, a man called Tracy recalled an NDE he had a child, when he was in a coma. In a similar way to David and Zak in the previous chapter, he described how he was "was no longer in my body. I had crossed over to a wondrous place of pure peace and love." He felt he could choose whether to return to his body or not and felt so serene that he wanted to stay where he was. However, then he saw a vision of "the future of my life if I returned . . . Everything! Meanings, events, pain, sorrow. Not just that, but why we are here! . . . How everything we do or don't do affects everything, everyone. How it all ripples through time and space." At the same time, Tracy was aware that nothing "was set in stone". Although in one sense he felt that the future was destined, in another it "was and was not".[17]

Can the Life Review Be Explained in Neurological Terms?

But hold on, you might argue, rather than entertaining such counterintuitive notions of "spatial time" and glimpsing the future, surely we should consider a much simpler option: that life reviews can be explained in neurological terms. Perhaps – in the same sceptical interpretation that is often used for NDEs – they are simply a neurological anomaly caused by physiological changes when a person is close to death. If this is the case, they have no more significance than a dream or a sequence of memories that we recall while lying in bed at night.

In the last chapter, we saw that there are many attempts to explain NDEs in standard scientific terms. In comparison, attempts to explain the life review are less substantial, both in number and detail. For example, the psychologist Susan Blackmore has tentatively suggested that the source of life review is the brain's limbic system, which is sensitive to the effects of anoxia (a lack of oxygen) and also linked to memory.[18] Along similar lines, after analysing a paltry seven accounts of life reviews, in 2017, a team of Israeli suggested that they may be caused by hypoxia (a lack of oxygen to the brain) and blood loss when a person is close to death.[19]

Another tentative theory was put forward in 2011 by the psychologists Dean Mobbs and Caroline Watt. They suggested that the life review in NDEs may be caused by a stimulation of the brain's noradrenaline system, which "has been shown to enhance and consolidate memory". This theory is based on an alleged link between NDEs and rapid-eye movement (REM).

In REM sleep, the brain's noradrenaline system is stimulated, so this could also be the case in NDEs. In addition, REM sleep is "thought to underlie the consolidation of memories".[20] However, the link between REM sleep and NDEs is very tenuous. Mobbs and Watt discuss just one case in which a person with diabetes displayed rapid-eye movement during an NDE. This theory depends on so many unproven links and assumptions that it carries very little weight.

In 2022, another theory about the life review attracted a lot of media attention. A team of scientists led Dr Ajmal Zemmar of the University of Louisville accidentally recorded the brain activity of an 87-year-old man as he died of a heart attack. EEG recordings indicated that his brain activity took around 30 seconds to fade and disappear after his heart stopped beating. All the man's brain waves decreased in activity over those 30 seconds, but his gamma waves decreased more slowly than others. There was also a surprising integration between different types of brain waves, with the "strongest coupling" between alpha and gamma waves. In their paper based on the case, the authors suggested that "given that cross-coupling between alpha and gamma activity is involved in cognitive processes and memory recall in healthy subjects, it is intriguing to speculate that such activity could support a last 'recall of life' that may take place in the near-death state."[21]

Although media articles about the research promoted it as an explanation for the "life flashing before your eyes" phenomenon, the researchers themselves were more circumspect. They admitted that it was problematic to draw general findings from just one case, especially as the man had a traumatic brain injury and was taking anti-convulsant medication (both of which affect brain waves). Another issue with the study is that memory recall involves so many different brain processes that to highlight a coupling of alpha and gamma activity seems like cherry-picking. For example, a more recent study has highlighted a strong link between theta waves and memory.[22] And as we saw above, other theories of the life review associate memory with the brain's noradrenaline and limbic systems.

In other words, all the above theories are based on highly tentative links to brain structures or neurological processes associated with memory. But even if these correlations were valid, another general weakness of the theories is that they portray the life review as simply a stream of memories. As we have seen, the events and experience of the life review are much richer and more vivid than ordinary memories. In addition, the life review is almost always described as a highly coherent, ordered experience, accompanied with feelings of serenity. As I pointed out in relation to near-death experiences, this is completely unlike the chaotic flood of images that normally takes place under cerebral anoxia or hypoxia, usually accompanied with feelings of anxiety. It's also dissimilar to the effects of increased noradrenaline, which is associated with stress and the "fight-or-flight" response.

Perhaps most importantly from our point of view, none of these theories explain the extraordinary temporal effects of NDEs. Perhaps anoxia or increased noradrenaline may trigger a rapid flow of memories, but how could they cause a review of every experience of a person's life – often in perfect sequence – from multiple perspectives, in the space of a few seconds? It also goes without saying that these neurological theories cannot explain the future visions of some life reviews.

As with NDEs, I believe that the key to understanding the life review is altered states of consciousness. The life review is not an altered state of consciousness in itself but a phenomenon that occurs *in* radically altered states of consciousness, such as the NDE or life-threatening incidents like falls or near-drownings. It isn't clear why some people experience a life review in these moments while others don't – as we saw earlier, they only occur in 20–25 per cent of NDEs

– but certainly, the experience is only possible in a radically altered state of consciousness induced by proximity to death.

As I also pointed out in relation to NDEs, if the life review is an *authentic* experience (rather than a neurological anomaly) it implies that linear time is a construct of our minds, rather than a fundamental feature of the world. After all, this is one of the significant insights that people gain from life reviews and NDEs – that, in Anita Moorjani's words, "our earthly minds convert what happens around us into a sequence; but in actuality . . . everything occurs simultaneously, whether past, present or future."[23]

We're going to explore this insight in more detail in the final chapters of this book.

CHAPTER 7
EXPLAINING TIME EXPANSION AND CESSATION

Altered States of Consciousness and Time

Before we plunge more deeply into the strangeness of spatial time, let's pause for a moment to take stock. In this chapter, I'd like to establish the connection between altered states of consciousness and TEEs and TCEs more deeply. I'll do this partly by examining other possible explanations for TEEs and also by looking at other altered states of consciousness in which time expands or disappears.

In the last two chapters, we discussed some attempts to explain near-death experiences and life reviews in neurological or physiological terms, finding such theories wanting. While reading the earlier chapters of this book, you might have wondered whether there are any attempts to explain TEEs or TCEs generally in such terms. In Chapter 2, we rejected the ideas that TEEs are hallucinations or an illusion of recollection. But perhaps there are other naturalistic explanations of TEEs or TCEs that are worth considering, so that we don't have to resort to esoteric notions of timelessness?

The Finnish philosopher Valtteri Arstila has suggested that emergency TEEs are due to increased levels of noradrenaline in the brain, related to the "fight-or-flight" response. (You may recall from the last chapter that noradrenaline has also been linked to the life review.)

Arstila argues that high levels of noradrenaline could account for characteristics of TEEs such as more focused attention, increased speed and accuracy of responses and improved clarity of thought.[1]

This theory seems to fit with the fact that many TEEs occur in life-threatening situations. However, the most common theme of TEEs (in accidents and other situations) is calmness and wellbeing, which doesn't fit with the fight-or-flight response or the effects of noradrenaline. The fight-or-flight response features anxiety and stress, both of which are strikingly absent from most TEEs. In addition, as we have seen, TEEs don't *just* occur in accidents and emergencies but also in many non-emergency situations such as sport, meditation, under the influence of psychedelics or while listening to music. Except sport, none of these are situations where one would expect to find high levels of noradrenaline. Meditation (and other relaxed states such as listening to classical music) bring a high level of stillness and wellbeing, in stark contrast to a fight-or-flight response.

A slightly different approach was taken by the neuro-scientist Bud Craig, whose ideas on time perception we discussed briefly in Chapter 2. As we saw then, Craig believed that time perception is linked to the insular cortex, the main part of the brain that registers bodily sensations and emotions. He suggested that radical time expansion may be due to the intense emotion of life-threatening experiences, such as (in his examples) accidents or parachute jumps. In these moments, what he calls "global emotional moments" – the entirety of our different feelings and sensations at one moment – pass through the insular cortex at a higher speed. This slows down our sense of time, perhaps even to the point where time apparently stands still. As Craig wrote, "The rate of passage of global emotional moments must effectively speed up during an intensely emotional period and time in the objective world would appear to 'stand still' to the subjective observer."[2]

However, this theory faces the same problems as Arstila's. Even in life-threatening situations, most TEEs or TCEs are not emotionally intense, but experiences of calm detachment and wellbeing. They also occur in non-emergency situations that lack any emotional intensity. For example, in deep meditation a person may be in a state of emotional stillness with very little awareness of physical sensations and yet experience a massive expansion of time.

In his book *Your Brain Is a Time Machine*, Dean Buonomano discusses the theory that TEEs are due to the speeding up (or "overclocking") of brain processes. He mentions three processes that could result in "overclocking": the speed at which electronic signals travel along axons (the "cables" inside our brain cells that carry signals from the main part of the cell), the speed at which signals travel through brain cells and the speed of voltage change in brain cells. Perhaps these processes speed up during life-threatening moments, creating a subjective impression that more time has passed. However, Bounomano points out that the speed of these processes is determined by biochemical and neurological processes that have little capacity for variation, and so can't account for radical time expansion. For example, he states that "changes in the firing latency of neurons are unlikely to amount to speed increases of more than 10 or 20 per cent."[3] So this obviously can't account for TEEs in which time speeds up by many orders of magnitude.

Buonomano also discusses the "hypermemory" theory that we tend to remember traumatic or dramatic events in more detail. This is essentially the same as the "retrospective illusion" theory that we examined in Chapter 2, which suggests that TEEs don't occur in the moment but only in retrospect, due to the enhanced memories that emergency situations create. In support of this theory, Buonomano notes that during intense or dangerous events our brains release neuromodulators that can enhance memory. However, he also points out that this theory doesn't explain reports of people thinking and acting more rapidly than normal or the "often compelling subjective sensation that the slow-motion effects occur in the moment".[4]

In Chapter 1 I pointed out that at present the neurological basis of time perception is unclear. In view of that, it's not surprising that we lack a clear idea of the neurological basis of TEEs. Perhaps if we could identify the neurological processes associated with time perception, then we could identify the neural basis of TEEs. If we keep conducting research and increasing our knowledge of the brain, we will eventually find the neural correlates of time perception and hence also the neurological causes of TEEs.

Consciousness and the Brain

Or perhaps not. In my view, some scientists and academics are far too ready to try to explain subjective experiences in neurological terms. Such scientists assume that human consciousness is a product of brain activity and so all types of conscious experience must be produced by specific types of brain activity. This assumption is responsible for the "medical model" of mental illness, which regards mental conditions like depression or psychosis as due to brain malfunctions, caused by chemical imbalances or too much or too little activity in different parts of the brain.

There are many problems with these assumptions, which I don't have space to fully explore here. (I examine them in detail in my book *Spiritual Science*.) After decades of intensive research, the attempt to find the "neural correlates of consciousness" has been a failure. In a similar way to NDEs, there are innumerable speculative theories about the different brain structures and processes involved in consciousness, with very little consensus.

Besides this, there are many strange mismatches between neurological activity and conscious experience. There are many notable cases of people who are born with massive amounts of normal brain matter missing – with up to 90 per cent fewer brain cells than normal – but who nevertheless have completely normal conscious experience, without any impairment.[5] A similar anomaly is "terminal lucidity", when dying people with severe brain damage – perhaps due to dementia, a stroke or meningitis – regain normal consciousness for a short time, becoming lucid and alert.[6] In other cases, people may have extremely minimal brain activity – such as in a deep coma – and yet report intense conscious experiences. In Chapter 5 we looked briefly at Eben Alexander's story, when he underwent a long series of intense conscious experiences despite a bare minimum of brain activity. And of course, near-death experiences represent the most extreme mismatch of all. During NDEs, consciousness becomes much more intense than normal in the apparent absence of *any* brain activity.

More generally, it is difficult to identify the "neural correlates" of any specific feeling or conscious experience. Although it is possible to associate different parts of the brain with different functions (such as vision, language or the regulation of emotions and sensations), specific feelings such as depression, happiness or love can't be reliably "mapped" to particular parts or processes of the brain. In the case of depression, for example, the simplistic belief that it is caused by a chemical imbalance – for example, a low level of serotonin – has now been discredited.[7] The consensus is now that depression can't be specifically located but is, in the words of one neuroscientist, "like other abnormalities of higher mental functions ... distributed across several brain regions".[8]

In view of this, perhaps we shouldn't expect ever to establish the neural correlates of TEEs or TCEs (or time perception itself). And even if we did, it wouldn't necessarily mean that we had found the *cause* of time expansion experiences. The correlates may only be traces of the experiences rather than their source, like footsteps in the snow that show (rather than cause) the movement of walkers. As psychology students are often told, correlation does not mean causation. This applies to all neurological theories of human experience, including NDEs and the life review. Even if scientists one day confirmed that the life review is associated with certain neurological processes, it wouldn't mean that the brain activity actually *causes* the life review. It could just be a trace or signal of the experience.

All of this relates to what the philosopher of consciousness David Chalmers called "the hard problem". It might seem natural to assume that the brain produces consciousness or that specific types of brain activity induce specific feelings or states, but the brain is just a physical object consisting of cells, molecules and atoms. It doesn't contain any traces of conscious experience. To say that this soggy lump of matter can "produce" consciousness is like suggesting that water can turn into wine. No matter how deeply you look into the brain, you don't come across a possible process by which this miracle could occur.

So my view is that we don't need to look for a neurological cause of – or even a neurological correlation for – TEEs or TCEs. It is enough to look for a *psychological* cause. The mind is not just a shadow of the brain. In the philosophical approach that I call "panspiritism" (described in *Spiritual Science*), consciousness is a fundamental quality that pervades all space and all objects. The role of the brain (or one of its roles) is to act like a radio transmitter and "pick up" universal consciousness, channelling it into our inner being. This means that the fundamental nature of the mind is not physical. The essence of the mind is universal consciousness, which is non-material. As a result, the activity of the mind doesn't have to be reduced to the brain. It has its own independent status, with its own patterns and laws and can be investigated independently.

TEEs as an Adaptive Trait

Before we leave standard scientific explanations behind, there is another possibility to explore: that TEEs are an evolutionary adaptation. Perhaps our ancestors developed them as a way of increasing their chances of survival in dangerous situations. This links to the field of evolutionary psychology, which suggests that modern traits can be traced back to prehistoric times, when they conferred some survival benefits for our ancestors. In genetic terms, the genes associated with traits were "selected" by evolution and passed down from generation to generation to present-day humans.

Picture what life was like for our early ancestors, surrounded by wild animals and dangerous natural phenomena. It would surely have been beneficial for them to slow down their experience of time in emergency situations. After all, as we have repeatedly seen, TEEs allow us to think and act rapidly, and to take preventative action that would otherwise be impossible. There is no doubt that they *are* an aid to survival.

So perhaps when we have TEEs, we are reliving a mental trick our ancestors developed to help them survive encounters with lions or bears, or flash floods or landslides. The anthropologist Edward T Hall provided a pertinent example of a TEE when he found himself face to face with an escaped mountain lion:

> My first awareness of what had happened was when I felt something brush by the calf of my right leg. Then, as I watched the lion lick a spot of grease next to my toe, time slowed down . . . Putting years of experience with animals to work, while I mentally reviewed and rejected a half dozen options and their scenarios, the only workable solution seemed to be to make friends.[9]

However, as with other explanations of emergency TEEs, this doesn't account for the other situations where

TEEs occur, such as during meditation, under the influence of psychedelics or while listening to music. As a paper by a group of time perception researchers led by Andrea Piovesan pointed out, "If lengthening of subjective duration is to be adaptive, it must also be limited to circumstances of specific threat."[10] As this is not the case, TEEs cannot be explained as an evolutionary adaptation. And in any case, even if this theory were true, it would only tell us *why* – not how – TEEs occur. We would still need to explain the psychological processes that cause them.

As a final point, although there is some logic to viewing TEEs as a survival aid, it is more difficult to see TCEs in those terms. What would be the survival benefit of transcending linear time and viewing all past and future events in parallel with the present? Surely that would hinder our chances of survival, rather than enhance them. TEEs and TCEs are so closely linked that any theory of time perception has to account for them both.

Altered States as the Key to TEEs

Let me admit that I don't have a completely clear and detailed explanation of TEEs or TCEs either. However, I've hopefully made it clear by now that the key to understanding them is altered states of consciousness. I will now explain this relationship in a more detailed and nuanced way.

In Chapter 2, we discussed the work of the early psychologist William James, whose ideas about time perception are still valid today. James had a very wide-ranging concept of psychology. Unlike many modern psychologists, he felt the field should encompass paranormal phenomena and mystical experiences. As well as writing one of the great studies of mystical experiences, *The Varieties of Religious Experience*, he self-experimented with psychoactive substances such as nitrous oxide and ether. As a result of his experiments,

James famously wrote that human beings' normal state of consciousness is "but one special type of consciousness, while all about it, parted from it by the flimsiest of screens, there lie potential forms of consciousness entirely different".[11]

James felt that we human beings often make the mistake of assuming that the world that our normal consciousness reveals is reality itself. He described this as prematurely "closing our account" with reality. In James's view, our normal consciousness is actually unreliable, in that it offers us a filtered and limited vision. On the other hand, different types of consciousness (such as mystical experiences) reveal a wider and more intense reality. (I made the same point in Chapter 5, in relation to NDEs.) As a result, they can provide insights about the nature of reality.

One insight that emerges from altered states is that time perception is not objective or absolute, and our normal experience of time perception has no special claim for validity. Time perception is a variable that depends on our state of consciousness. Or to put it another way, as many people become aware during NDEs and life reviews, our normal experience of time is a *construct*. It is created by the psychological processes and structures of our normal state of consciousness – our thought processes, perception, sense of identity, sense of self-boundary or separation, and so on. When these processes and structures are functioning normally, we experience time in a normal, fast-flowing linear way. But when, in intense altered states, these processes change significantly – for example, when we lose our normal sense of identity and separation, and our perception becomes more intense – our experience of time changes significantly too. Then we enter a different timeworld.

Over the last few chapters, we have established the causal link between altered states and TEEs or TCEs very firmly, from multiple perspectives. This is the common factor in all the different situations where TEEs occur, including accidents, sports, meditation, psychedelics, NDEs and the life review – that is, they are all altered states of consciousness. We've also seen that this causal link explains why some TEEs are more intense than others. Mild altered states of consciousness cause mild TEEs, while intense altered states (such as NDEs or psychedelic experiences) bring intense TEEs. Here there is a link to TCEs too. TCEs *only* occur in very intense altered states, such as deep mystical experiences or NDEs. In my view, TEEs and TCEs are not essentially distinct, but part of a continuum. There is a point of intensity at which a TEE becomes a TCE, in the same way that there is a certain point at which a slowing vehicle comes to rest.

In addition, the link between altered states and TEEs explains why some people are more prone to TEEs than others – for example, why only one out of three or four passengers in a car may have a TEE in an accident, or why one person may have several TEEs in their life while someone else may never have a single one. This is because some people are more susceptible to altered states than others, with thinner self-boundaries that make them open to unusual experiences.

A final link is that altered states help to explain extra-ordinary sporting prowess. As we saw in Chapter 3, the difference between the most extraordinary athletes and their less successful peers may be that they have easy access to altered states, and so can slow down time almost at will.

Transcendence of Boundaries

Perhaps we could be more specific though. What aspect of altered states is most significant in TEEs and TCEs?

Here let me return to the term "self-system", which I used for the fourth law of psychological time ("Time passes very slowly in intense altered states of consciousness, when our normal psychological structures and processes are significantly disrupted and our normal 'self-system' dissolves.") As I noted then, this term refers to our normal sense of identity, and all the normal psychological processes and functions that constitute it.

One of the main features of our self-system is its strong boundary. We normally experience ourselves as mental entities – or subjective selves – who inhabit our brains and bodies. We feel as if we live "in here", while the rest of the world appears to be "out there". We are all islands, enclosed within our own mental space. Our normal sense of linear time is closely linked to this sense of separateness. In the same way that the world appears to be separate, we perceive time as separate to us. It seems to flow by independently of us, like a river that flows by at the bottom of our garden.

As we tend to do with our normal consciousness, most of us perceive separateness as a given, a fundamental reality of life. However, all the world's spiritual traditions – from Buddhism to Vedanta to Daoism and the mystical paths of Christianity and Islam – are based on the principle that separation is illusory. The goal of all these paths is to transcend our normal sense of separation. True wellbeing only arises once we have transcended separation. From time to time, we transcend separation on a temporary basis, too. As I noted in Chapter 5, this is one of the main features of awakening experiences.

In intense altered states, the normal functioning of the self-system is disrupted. In especially intense altered states, the self-system may even dissolve away entirely. In these moments, our normal self-boundary either softens or disappears. This is the key feature of psychedelics, at least in strong doses: they bring about "ego-dissolution". (Researchers have developed an "ego-dissolution inventory" as a tool to study psychedelic experiences.) In my view, the same phenomenon occurs in life-threatening accidents and emergencies, in deep meditation or when athletes enter the highest level of the Zone. In all these experiences, separation fades away and our normal sense of time expands. The more separation fades, the more time expands. And when the normal self-system fades away completely, time ceases altogether. We step out of linear time into a world of spatial time.

I'm not the first person to suggest this link between the self-system and time. In his book *Felt Time*, Marc Wittmann notes that the sense of time and the sense of self are inseparable. As he puts it, "In extraordinary moments of consciousness – shock, meditative states and sudden mystical revelations, out-of-body experiences or drug intoxication – our senses of time and self are altered; we may even feel time and self dissolving ... Without a concept of self, time does not exist."[12] In other words, the dissolution of self is accompanied by the dissolution of time. Or as the psychoanalyst Peter Hartocollis explained, our sense of time is indistinguishable from "the experience of the self as an enduring, unitary entity that is constantly becoming".[13]

This makes sense when we consider how our sense of time develops in early life. Our sense of separation and our sense of time develop in parallel. In the first months of their lives, babies have neither a sense of separation, nor a sense of time. They can't tell separate objects or themselves and objects apart. They are fused together with the world; they can't sense where they end and where it begins. They also experience timelessness. In the same way that they can't distinguish between objects, they can't separate one moment from the next.

Slowly, as our sense of self and of separation develops, we emerge from this timeless realm. We start to become aware of ourselves as separate entities, apart from the world. We become aware of the separation between different objects. And in parallel, we become aware of separation between events. We develop a sense of sequential time, a sense of the past and the future, which is encouraged by the development of language, with its past, present and future tenses. During early childhood, our self-boundary is weak, which is one reason why time passes so slowly. (The other reason – as discussed in Chapter 1 – is that children's perception is so vivid and fresh, which brings more information processing.) As our self-boundary grows stronger – and as our perception grows more automatic – time speeds up. In other words, the more separation we experience, the faster time seems to pass.

In other words, the main cause of TEEs and TCEs is the dissolution of the boundary of the self-system. When the boundary between us and the world disappears, so does time. It's important to note that this just simply means losing self-awareness. As Peter Hartocollis indicates above, the self as an *enduring entity* has to fade away. There are many situations when we lose self-awareness, without experiencing ego-dissolution. This is what happens in flow states – although we lose self-awareness, the self-system remains intact. This is why, as we have seen, flow states don't cause time expansion – just the opposite; they cause time contraction due to reduced information processing.

Time and Space

In view of the connection between TEEs and the transcendence of boundaries, it's significant (though not surprising) that TEEs and TCEs often bring changes to *spatial perception*. We've seen this throughout our examination of TEEs and TCEs. In emergency and sporting TEEs, people sometimes report a sense of being outside their bodies, watching themselves from a distance or observing the situation in overview. In Chapter 2, we heard from a man who felt "like I was next to my physical self" while falling downstairs, and a woman who sensed that she wasn't "even in my own body" while saving her children from a fire. In Chapter 3, Billie Jean King reported that she played her best tennis feeling like "an observer in the next room", while racing driver Mika Häkkinen felt like a "bird of prey" watching over his car. As we saw in Chapter 5, near-death experiences almost always feature out-of-body experiences. Typically, these occur at the beginning of the experience, when people are surprised to find themselves looking down on their bodies from above, perhaps observing their own medical procedure.

Another significant change to spatial awareness that occurs in NDEs is a sense of *expansion*. A person's identity extends beyond their mind and body, and merges with other people or the whole world. As David Ditchfield reported after his accident at the train station (see page 105), "my awareness felt massively expanded, a part of every single thing in my surroundings, as though my awareness was not just looking out through my eyes but was part of an infinite field of awareness . . . I was this eternal field, and it was me. We were one and the same."[14] As we saw in Chapter 4, this is also one of the essential features of awakening experiences: a transcendence of separateness, a sense that we are no longer isolated entities but part of a unity of being, one with other people, with nature and the whole cosmos.

In other words, in TEEs and TCEs, our physical boundaries are revealed as illusory, along with the boundaries of linear time. Once the self-boundary fades, we are no longer trapped inside our minds and bodies. This doesn't mean that we cease to exist as individuals or return to the psychological state of early childhood. Our identity is still associated with our body and mind. But we are no longer separate or limited, in space or time. It's entirely possible to function as mature integrated personalities, dealing with the practical business of our daily lives, while experiencing a sense of oneness and expansive time. This is what spiritual traditions such as Hindu Vedanta or Taoism teach us: that transcending the illusion of separation leads to a much richer and more fulfilling life. (We'll look at this in more detail in the last chapter of this book.)

The idea that time and space are interdependent links to the ideas of the 18th-century German philosopher Immanuel Kant, who described both space and time as constructs of the human mind. At a level of fundamental reality – which Kant called the *noumenon* – neither space nor time exist. The human mind imposes them both on the world to make sense of our experience. In this way, time and space are inextricably linked, and can only be understood in terms of each other.

Modern physics also treats time and space as interdependent. Whereas Isaac Newton viewed time and space as separate features of the universe, the Austrian physicist Hermann Minkowski (who was Einstein's tutor) realized that Einstein's Theory of Relativity logically implied that time and space were inseparable. They couldn't be investigated independently, only in unison, as "space–time". (We will discuss Minkowski's and Kant's theories in more detail in chapters 9 and 10 respectively.)

In summary, our normal fast-flowing linear experience of time is a construct, produced by the strong boundary of our normal self-system. When that boundary fades or disappears, in intense altered states of consciousness, then time itself fades and disappears.

Time Contraction in Altered States

One argument against my theory might be that altered states don't *always* cause a slowing down of time. As we've seen, the flow state speeds up the passage of time, as often does meditation. Time also tends to pass quickly during therapy and counselling sessions, when clients may experience a mild altered state of consciousness due to relaxation or the intensity of their connection with their therapist. This is also usually the case with hypnotism (as we will see shortly).

However, the important point is that these are only *mild* altered states. They don't affect the structure of the self-system to any significant degree. There is merely a time-quickening effect due to reduced information processing. You could compare it to living in a tent. In mild altered states, we change the interior of the tent – perhaps we buy a new camping mat or ground sheet or move our cooking utensils to a different position. In other words, the self-system remains intact. But in intense altered states, the frame of the tent is dismantled, its covering flies away and we find ourselves in the open air.

This relates to our earlier discussion – in Chapter 5 – about drugs that usually have a time-contracting effect, such as cocaine, heroin and alcohol. These drugs certainly induce altered states of consciousness, but not ego-dissolution. They produce pleasant biochemical changes and bring some changes to aspects of consciousness such as perception, attention or information processing. But they don't cause any fundamental changes to the self-system, which remains intact. In the metaphor above, the tent remains upright, even if its interior changes.

Time in Hypnosis

There is one altered state of consciousness that we should discuss in a little more detail, as it has such varied and striking effects on time perception: hypnosis.

About 15 years ago, I went to see a hypnotherapist. For many years, I had anxiety about flying and I wanted to deal with the issue before a transatlantic flight. I had tried hypnosis once before, but it didn't work very well. Back then, I didn't seem to be amenable to the hypnotist's suggestions and barely registered any change in consciousness. But this time, I quickly fell into a state of deep relaxation. I felt like I was floating along a river on a hot summer's day. I was still aware of my surroundings and felt that I still had control of my own mind and body (although this could have been illusory, of course) but I certainly entered a mildly altered state of consciousness. I was shocked when the therapist guided me back to a normal state and told me the session was over. The session lasted for 90 minutes, but to me it seemed as if about

EXPLAINING TIME EXPANSION AND CESSATION

40 minutes had passed. (The therapy did help to reduce my flying anxiety by the way!)

This is a standard experience. Most hypnotized subjects underestimate time, usually (as I did) by 40–50 per cent. This makes sense in view of the low level of information processing that occurs under hypnosis. It's not dissimilar to flow or a good meditation. We close our senses to the external world, narrowing our attention down to the hypnotist's instructions. As we become relaxed and focus our attention, our minds slow down. The stream of thought may even stop altogether. But this quickening of time suggests that, as with flow, hypnosis doesn't normally bring about a dissolution of the normal self-system.

Because of this time-quickening effect, some hypno-therapists and NLP (neuro-linguistic programming) practition-ers recommend hypnosis as a way of dealing with unpleasant situations, by making them pass quicker. For example, the hypnotherapist Paul Gustafson has recommended hypnosis for those of us who are afraid of the dentist. As he wrote, "Hypnosis can help dentistry clients as an effective analgesic adjunct. It relieves anticipatory anxiety [and] distorts time perception, speeding up the procedure."[15]

This isn't the whole story though. Under hypnosis, time can become incredibly flexible, depending on the suggestions of the hypnotist, and the depth of the hypnotic trance. The pioneers of research in this area were the American psychologists Winn Cooper and Milton Erickson. In 1954, they published *Time Distortion in Hypnosis*, describing their time-slowing experiments with hypnotized subjects. One of their methods was to pretend to slow down a metronome until their subjects believed it was beating just once every minute. With good subjects who entered a state of deep trance, time slowed down by similar orders of magnitude to emergency TEEs or even NDEs.

To ensure that people weren't simply imagining that time had slowed down, Cooper and Erickson asked their subjects to accomplish mental tasks, such as preparing lectures, designing dresses or planning complicated meals. People accomplished these tasks in a matter of seconds, while reporting that they felt that long periods of time had passed. Most significantly, they reported intricate details of their thought processes, and even made drawings and notes. For example, a woman whose hobby was designing dresses drew intricate illustrations of her designs, with detailed instructions (reproduced in Cooper and Erickson's book). She experienced a period of a few seconds as if it was several hours.

The important point is that this radical time expansion occurs in a state of *deep* hypnotic trance. It's similar to a deep meditation that brings time expansion (or cessation), or in sport, when super-absorption takes an athlete into the Zone. Normally, under hypnosis, the self-system remains intact. But in a deep trance, the self-system dissolves away and we may enter the same expansive timeworld as the Zone and deep meditation. In this regard, it's significant that Milton Erickson was renowned as one of the greatest hypnotists of his time, and that he and Cooper took care to work with people who were highly susceptible to hypnosis.

Disturbances of Self and Time

All the altered or transliminal states we've examined so far in this book have been positive experiences. Despite the critical events that sometimes generate them, emergency situations, the Zone experiences of athletes, awakening experiences (including psychedelic experiences), NDEs and the life review are all states of wellbeing and all have positive transformational after-effects. This applies to hypnosis too, which is usually a deeply relaxing state.

However, some transliminal states (that is, states in which people lose their self-boundary) are extremely negative. In some situations, when the self-system dissolves, people may experience mental disturbance, instability and confusion. When this happens, time expansion or cessation become a painful experience, adding to a person's overall distress.

This can happen in schizophrenia or during episodes of psychosis. Schizophrenia is difficult to define precisely – it's a general term to describe a range of symptoms that occur when our self-system doesn't function coherently or when it breaks down altogether, including (significantly) a loss of the sense of a boundary between oneself and others. Symptoms may include hallucinations, disorganized thinking, the inability to distinguish one's own thoughts from reality and a feeling of disconnection from one's own body or reality itself. However, in recent years, schizophrenia has been interpreted more frequently as a *temporal* disorder. In the words of one group of researchers, it is "associated with a fundamental disturbance in the temporal coordination of information processing in the brain".[16]

People who are diagnosed with schizophrenia find the normal human timeworld very difficult to navigate. Time is an alien concept, which they struggle to make sense of. They often have difficulty estimating how much time has passed, or distinguishing between events that have already happened and events that they expect to happen. The normal smooth flow of time seems to be disrupted, as if moments have become detached from one another. Many laboratory studies have found that people with schizophrenia struggle to accurately judge the length of intervals, or to accurately reproduce or guess time periods. As another group of researchers put it, "temporality may lose all organization and meaning."[17]

There may also be a sense that time is passing incredibly slowly. One tragic case was described by the psychiatrist Dr A Moneim El-Meligi. His patient was a severely depressed young man who suffered from sensory distortions. Even simple actions like lifting a cup or drinking a glass of water were extremely difficult, as he would see the cup or glass growing larger and larger. As he told El-Meligi, "Time is the worst thing. It seems so long that a month period is unimaginable. Time seems endless . . . You just cannot see the end of it. It seems so long. I hate time, it seems very real. I make myself sleep in order to conquer time, but it does not work."[18] The patient described how his slow time perception meant he couldn't follow TV programmes or enjoy baseball matches, as they seem interminable.

In other severe cases of schizophrenia, people may report a feeling that time has frozen, stopped flowing or ceased to exist. Patients may feel as if life is a series of disconnected snapshots with no relation to the future or the past. In the words of the participants of one study, the world is "like a series of photographs". Other participants reported that "I lost [my] sense of time" and "time doesn't mean nothing to me."[19]

All of this largely applies to dementia or Alzheimer's too. Like schizophrenia, dementia can be seen as a disorder of time perception. People with dementia lose their awareness of time and the ability to measure or estimate it. They seem to step out of the present into an indeterminate realm of different time perspectives where they feel disoriented. In a fascinating book called *Reconceptualising Dementia,* Dr Mina Drever suggests that the primary characteristic of dementia is a lack of mindfulness, or an inability to attend to experience of the moment. People with dementia or Alzheimer's live outside time, often regressing to different points of their own lives – often their childhood or youth – and referring to visitors as if they are people from their past. Drever observed this with her own mother, for whom "The past and the present became one . . . Time and space were unified in the reality of remembered episodes, bursts of memories that presented themselves to her mind in cinematic snapshots."[20] This is the main reason why people with dementia can't focus on a mental task or on an activity for more than a few moments: their attention is constantly being pulled away to different times and places. As Drever writes of another person living with dementia, "She is not being mindful, because her mind is covering the whole spectrum of her life's experience."[21] As a result, Drever uses the term "amelesia" (literally, unmindfulness) instead of dementia.

As I would interpret it, people with dementia or Alzheimer's (like people with schizophrenia) don't have a stable self-system with a strong boundary. As a result, their minds can't create a stable and coherent timeworld. We've seen that our normal sense of time is a construct, but it is not a given that the construct will be created. If a person's psychological processes are impaired, then they may not be able to construct a normal sense of time.

The question of why some transliminal states are positive while others are negative is a complex one. Why do people with schizophrenia and dementia experience distress, whereas transliminal states such as NDEs and emergency TEEs feature intense wellbeing and even bliss? Why do people with schizophrenia and dementia struggle to cope with life, whereas those who experience ongoing wakefulness (or enlightenment) shift to a higher-functioning level?

The important point is that in mental disorders the normal self-system is simply disrupted or broken down, resulting in chaos and distress. But in positive transliminal states, the normal self-system is *transcended*, without any disorder or disruption. Psychological processes don't break down but shift to a higher level of functioning, becoming more integrated. The normal self-system is replaced by another, high-functioning self-system, which doesn't have a strong boundary and doesn't create a sense of fast-flowing linear time. In this state, a person doesn't feel overwhelmed or disoriented due to fractured time perception; they feel completely comfortable in their new expansive timeworld.

A strong self-boundary can help us to function in the world, but it is also possible to function without one. In fact, so long as our psychological processes are stable and coherent, we can live in a much more fulfilling and contented way with a softer self-boundary, in a state of connection rather than separation. As I noted in Chapter 5, this is the essential insight of the world's spiritual traditions: that separation creates suffering, and true contentment arises from sensing our fundamental connection – and ultimately oneness – with the world. As we also saw in Chapter 5, connection and wellbeing are the essential characteristics of awakening experiences. Some people may even experience these qualities on an ongoing basis, as a state of "wakefulness."[22]

We will return to these ideas in the final chapter of this book, when we examine the possibility of consciously inducing TEEs, and living in an ongoing state of time expansion. Before then, we need to look further into the notion that our normal sense of linear time is a construct – and even an illusion.

CHAPTER 8
THE FUTURE
IS WRITTEN

Precognition and Retrocognition

In 1988, a 28-year-old woman named Elizabeth Krohn stepped out of her car with her children and started to walk across the parking lot of a synagogue, to attend a service honouring her deceased grandfather. A thunderstorm was raging, and while her eldest son ran ahead, Krohn covered her youngest son and herself with an umbrella, preparing to walk the distance. After just a few steps, the umbrella was struck by lightning. As Krohn described it, "The power of the lightning strike was unlike anything I had ever felt. The deafening explosion, blinding light and crackling energy hit me all at once. The thunderous noise literally split our eardrums."[1]

Straight away Krohn found herself out of her body, looking down on the scene. She witnessed her children screaming and running toward the synagogue lobby, while a man ran toward them to find out what was wrong. At that point, she wondered why no one was looking at or speaking to her, then turned her gaze back toward the parking lot to see the smouldering remains of the umbrella and her own body lying inert on the pavement. Suddenly it struck her that she was dead.

After this, Krohn's experience followed the typical pattern of the near-death experience. She sensed a golden radiance toward her upper right and knew that she was meant to follow it. It led her to a place she later called "the garden", full of strange plants that "continuously blossomed into magnificent flowers that seemed to explode with colours from another spectrum inaccessible here".[2] The atmosphere of unconditional love that pervaded the garden is almost always a feature of near-death experiences. Soon she heard the familiar voice of her grandfather, although she felt that it was a higher presence or a guide – perhaps even God – that was speaking through him, rather than his own personality.

As with most NDEs, the garden was a place where time was not linear, where in Krohn's words, time was "perpetual . . . events and sensations all occur at once. This idea of simultaneous time, the physics of it, is something I understood while I was in the Garden but have difficulty explaining or even understanding now."[3] As if to emphasize this, she saw visions of future events, such as George Bush winning the next US election and the Cincinnati Bengals playing in the 1989 Super Bowl. Later, she wondered why she was shown such trivial pieces of information, but concluded that they were meant as future reminders of the non-linear nature of time. She was told by her guide that she could choose whether to return to her body or not. If she returned, she would have a third child, a daughter. However, her marriage would not endure the radical changes to her personality caused by this experience.

Despite the incredible beauty around her and the deep sense of peace she felt, Elizabeth knew she had to return for the sake of her children. When she came to she was still lying on the ground, with people running toward her from the synagogue. In terms of linear time, she felt that she been in the garden for two weeks, so it was a shock to find that only around two minutes had passed. As she reported, "I couldn't understand how I had received so much information and had been so completely transformed in such a short time. It was jarring and bewildering."[4] (As noted in Chapter 5, this is an even more extreme degree of time expansion than in Zak's or Eben Alexander's NDEs.)

In the aftermath of her NDE, Krohn began to have dreams of future events. Mostly, the dreams related to fairly trivial everyday events, but one night, in 1996, she had a nightmare about a plane crash. She could see "WA" on the wreckage and knew that all 230 passengers lost their lives. She was also aware that it was flight number 800. She was so troubled by the nightmare that she telephoned her mother and shared the details with her. The next day she turned on the TV to learn the tragic news that TWA Flight 800 had crashed in the Atlantic Ocean with 230 people on board and no survivors.

Later, in 2008, it occurred to Krohn to write email accounts of her dreams and send them to herself, to provide evidence of dates. On 15 January 2009, she was vacationing in Israel with her second husband and took an afternoon nap, during which she had a nightmare about another plane crash. She saw a plane floating on water, with people standing on its wings. At 2.57 pm Israel Standard Time (7.57 am Eastern Standard) she sent an email to herself that read: "Mid-size commercial passenger jet (80–150 people) crashes in NYC. Maybe in river. Not Continental Airlines. Not American Airlines. It is an American carrier like Southwest or US Airways."[5] Later that day, at 3.31 pm Eastern Time, Captain "Sully" Sullenberger landed a US Airways plane on the Hudson River, with no fatalities.

The Possibility of Precognition

Is it really possible that Krohn developed the ability to see the future – or precognition, as it is normally termed – as a part of the transformation caused by her NDE?

We have already touched on precognition in relation to NDEs and the life review. In Chapter 5, I briefly described Mary Neal's NDE, in which she learned that her son would die at a young age. We also discussed the future visions that Kenneth Ring describes as "flashforwards". In Chapter 6, we looked at life reviews containing future visions, such as when Tony Kofi saw himself playing the saxophone. As I pointed out in those chapters, the ability to glimpse future events makes sense in terms of a spatial view of time, in which the future is spread out before us, along with the present and the past.

This is why precognition is so relevant to this book: it implies that linear time is illusory, and that in some sense the future is already in existence.

Reports of precognition are more common than you might think. Glimpses of the future certainly don't just happen to people who are close to death or in the aftermath of NDEs. In the US, a 2018 survey found that slightly less than a half of the respondents reported an experience of knowing "something about the future that you had no normal way to know" (in other words, precognition), while over 40 per cent reported that they had received important information through their dreams.[6] Meanwhile, a 2022 survey of over 1,000 Brazilian people found that 70 per cent believed they had had a precognitive dream at least once.[7] (Perhaps this higher figure is due to a cultural climate of greater openness to anomalous experiences in Brazil.) As these statistics indicate, it is most common for precognition to occur in dreams. According to the pioneering paranormal researcher J B Rhine, this accounts for around 75 per cent of reports of precognition.[8]

Precognition appears to be common in children. In a Japanese survey, 42 per cent of university students reported at least one precognitive experience, with most stating that the experiences had begun between the age of six and ten.[9] According to one of the UK's leading researchers of children's anomalous experiences, Dr Kate Adams, precognition is so normal to some children that they think little of it, not regarding it as a special ability. She reports one example from a young boy called Ahmed, who had a dream in which he "was on my bike going down the hill and the bike was going really fast. Then suddenly straight ahead of me was a tree. I was going to crash into it then I swerved and just missed it! And the next day it really happened." When he saw the tree in front of him, he recalled the dream and steered away from the tree, avoiding injury. As a Muslim, he believed that the dream came from Allah, who was protecting him. As he told Adams, "Only Allah knows the future."[10]

Of course, this all depends on whether reports of precognition are genuine. Perhaps, as some sceptics claim, they can be explained away as coincidence, embellishment or self-delusion. This is the central question of this chapter: is there enough evidence to conclude that precognition really exists or are sceptics justified in explaining it away?

Disaster Precognition

One of the puzzling things about precognition is that it seems to operate mainly through both highly dramatic and completely trivial events. An early psi researcher, Herbert Saltmarsh, examined 281 strong cases of precognition collected by the Society for Psychical Research, and found that the most common variety was "natural or accidental death" with 99 cases. The only other common category was "completely trivial incidents", with 70 cases.[11]

It makes sense that fatal or tragic incidents are associated with precognition, due to their major emotional impact. After all, when we contemplate our past lives, these are the types of events we remember most clearly, so this could apply to future events too. As with Elizabeth Krohn, some of the most striking and convincing examples of precognition are linked to natural disasters, such as plane crashes, earthquakes and terrorist incidents.

In many cases, major disasters are anticipated by multiple people. One of the most famous examples is the Aberfan mine disaster of October 1966, when an avalanche of coal waste slid downhill after days of heavy rain, smothering part of the Welsh village of Aberfan, including a school – 144 people died, including 116 children. The day after the disaster, a psychiatrist called John Barker, based 100 miles (160km) away in Shrewsbury, arrived in the village to offer help to survivors and bereaved parents. Over the following few days, as he talked to witnesses, Barker heard several accounts of precognitive glimpses of the disaster. For example, the mother of one of the deceased children found a drawing made by her son the night before, showing crowds of people digging on a dark hill, under the words "The End". The day before the disaster a ten-year-old girl told her mother about a dream in which "I went to school and there was no school there. Something black had come down over it!"[12]

Barker was so struck by these reports that he decided to find out if people further afield had had premonitions of the disaster. Through a journalist acquaintance, he placed an appeal in the London Evening Standard. He received 60 reports that he deemed worthy of further investigation. In 22 of the cases, there was evidence that the premonitions occurred before news of the disaster broke. For example, a woman called Constance Miller described a vision she had while at a spiritualist meeting the night before the disaster. She saw an avalanche of coal sliding rapidly downhill, along with the terrified face of a boy at the bottom of the hill. She also had a vision of rescue operations. The experience was so uncanny that she told six witnesses about it that evening, who verified her story with Barker.

Intrigued by such stories, two months later Barker set up a "premonitions bureau" in London to collect and investigate other reports. In the first year, the Premonitions Bureau received 469 reports of future disasters or other world events. Most didn't seem pertinent – although of course, the open-ended nature of the future meant that such reports couldn't be dismissed outright – but several turned out to be uncannily accurate. In particular, Barker soon noted that there two respondents who regularly relayed detailed reports of imminent world events.

One was a middle-aged teacher, Lorna Middleton, who also reported a premonition of the Aberfan disaster. Around six months later, on 23 April 1967, she contacted Barker to report a vision of an astronaut travelling to the moon, a journey that would "end in tragedy". She saw the astronaut looking "petrified, terrified and just frightened".[13] Earlier that day, a Soviet astronaut had blasted off from Kazakhstan, in the USSR's first piloted space mission for over two years. Although the Soviet News Agency had released news of the mission that morning, it had not yet been broadcast by the British media, so it is unlikely that Middleton was aware of it. Even if she was, the details of the Soviet report were very vague. Tragically, the next day, the cosmonaut Vladimir Komarov died on his return journey to Earth, when his parachute failed and his capsule crashed into the ground.

The other uncannily accurate respondent was a Post Office switchboard operator called Alan Hencher. He first telephoned the bureau in March 1967 to describe a major air disaster that he felt was imminent. He reported a vision of a passenger plane "coming over mountains. It is going to radio it is in trouble. Then it will cut out – nothing."[14] He said that there would be 123 or 124 people on board and that all but one would die straight away. Around four weeks later, the wing of a passenger plane clipped the side of a hill during a thunderstorm while attempting to land in Cyprus. The next day, the *Evening Standard*'s headline was "124 die in airliner", although two more people died later in hospital.

Just ten days after this crash, Hencher reported a premonition of another air disaster. He predicted that a plane with "sweeping tail fins" would crash in the near future, with more than 60 fatalities, although he wasn't sure of the location. He specified that there would be some "miraculous escapes and a number of survivors" and that there "may be a lot of children involved".[15] Nearly five weeks later, what is still the fourth-worst ever aviation disaster in British history occurred, when a plane carrying British holidaymakers home from Majorca crashed close to landing at Manchester airport. Of the 84 people on board, 72 were killed. Many of the passengers were families with young children, while the survivors were freed from the wreckage by brave local residents and policemen. As Hencher predicted, the airplane had unusually large and wide tail fins.

1967 was an unusually bad year for transport disasters in the UK. On 1 November, Lorna Middleton wrote to the bureau, "I see a crash . . . maybe on a railway . . . people waiting in the station and the words Charing Cross. The sound of A CRASH."[16] Just four days later, the Hastings to London Charing Cross train derailed near a rail depot eight miles from its destination, killing 49 and injuring another 78. (As a point of interest, Bee Gee Robin Gibb happened to be on the train.) Alan Hencher didn't predict this crash, but he apparently sensed it as it happened. Working on the Post Office switchboard at the time, he developed a severe headache and had to leave his post. An hour after the accident – of which he knew nothing, as news travelled very slowly at that time – he wrote a note saying that he was sure that a major railway accident had occurred an hour ago.

Hencher and Middleton made other accurate predictions, including, sadly, the premature death of John Barker himself. Hencher warned Barker regularly that his life was in danger and in February 1968, Middleton had a vision of Barker standing with her own deceased parents, noting that "my parents were trying to tell me something." Barker was perturbed by their warnings but stoical. "I suppose anybody who plays about with precognition in this way to some extent sticks his neck out and must accept what he gets," he wrote.[17] In August 1968, Barker died of a brain haemorrhage, aged just 44.

The Gift of Precognition

A sceptic might argue that the apparent accuracy of Middleton and Henchers' reports is offset by hundreds of other predictions that didn't come true. But this isn't the point. To prove that an ability exists, we don't need to prove that everyone possesses it. It doesn't matter if some people claim to possess it but are shown to be self-deluded or fraudulent. As William James once famously stated, "To upset the conclusion that all crows are black, there is no need to seek demonstration that no crows are black; it is sufficient to produce one white crow; a single one is sufficient."[18]

The best way to think of precognition is as a gift. Although many people have occasional precognitive experiences – and although, as we'll see later, scientific experiments with the general population have consistently shown the ability to predict images and other stimuli at levels higher than chance – it is rare for people to predict future events reliably and regularly. Precognition isn't like the ability to cook or to drive, a skill that we can use at will on a day-to-day basis. It is an unusual talent, like composing music or writing poetry, which only a few people are truly good at and can consistently demonstrate.

Fortunately, just as there are a good number of composers and poets, there are a good number of "white crows" to demonstrate that precognition exists. I have written elsewhere (in *Spiritual Science*) about another remarkable case of disaster precognition – perhaps the most remarkable of all. In the early 1990s, a British artist and lecturer named David Mandell started to have dreams of disasters, which he took care to record and date. As an artist, Mandell was able to paint scenes from his dreams. He also included notes of some of the specific details – including locations and names – that came to him. Once he had completed the paintings, he photographed them at his local bank, underneath an electronic clock that displayed both the date and the year. (This was before the age of the Internet, so he wouldn't have been able to send emails to himself, as Elizabeth Krohn did.) Mandell anticipated a wide range of events, including the terrorist attack in a subway station in Tokyo in 1995; the attack on a primary school in Dunblane, Scotland in 1996; the Concorde crash in Paris in 2000; and the 9/11 terrorist attack. In the latter example, on 11 September 1996, he photographed a painting that showed two burning towers falling into one another. He also included an outline of the head of the Statue of Liberty and the silhouette of an aircraft flying downward.

There was experimental verification of Mandell's claims. He passed a lie detector test and experts who examined the negatives of his photographs found no signs of tampering. In an experiment devised by the sceptic investigator Chris French, 20 people were shown Mandell's paintings and given his interpretations along with alternative ones. They were asked to choose which interpretation best suited the paintings and in 31 out of 40 pictures, all 20 people chose Mandell's interpretation. French stated that this result was statistically highly significant and admitted that Mandell's claims "merit serious consideration".[19]

Premonition of the Date of Death

The 18th-century Swedish scientist and spiritualist Emanuel Swedenborg was a renaissance man with a bewildering array of skills and interests, who anticipated many of the key inventions and ideas of the modern era. During the first part of his life, Swedenborg wrote scientific treatises on astronomy, chemistry, geology and anatomy. In just the one area of anatomy, his insights and discoveries included the existence of cerebro-spinal fluid and its circulation through the body, the intricacies of the cerebral cortex, the existence of brain cells and the importance of the pituitary gland to neurological function. He was also an inventor who forward plans for early versions of a submarine, an aircraft, a car and even a machine gun. In addition, he wrote many philosophical works in which he was similarly prescient and prophetic.

However, midway through his life Swedenborg began to experience trance-states in which he believed he conversed with angels and was shown visions of heaven and hell. Unsurprisingly, his scientific contemporaries thought he had gone insane. However, Swedenborg remained outwardly stable and rational, and documented his explorations of the spirit world with the same detached precision as his scientific tracts.

In 1772, the British evangelist and theologian John Wesley received a letter from Swedenborg, who was living in London at the time. Swedenborg wrote, "I have been informed in the spiritual world that you have a strong desire to converse with me; I shall be happy to see you."[20] Wesley was shocked, as he did want to meet Swedenborg but had told no one about it. In response, he told Swedenborg that he was about to leave on a six-month speaking tour, but would be happy to meet him afterwards. Swedenborg replied that this would unfortunately be too late, as he was due to pass away the following month, on 29 March. And Swedenborg did pass away on that date. His servant also confirmed that he had predicted the date of his death.

The above story has special resonance to me, as my father also apparently predicted the date of his death. My father died in 2019 at the age of 79. He had Parkinson's disease and was quite weak and a little confused, but wasn't expected to die at that particular time. However, a week before his death, he told a friend, "I've found out when I'm going to die – it'll be a week today, next Thursday."

His friend was shocked but as my father didn't seem seriously ill, he didn't pay heed to his comments. My dad had just woken up from a nap, so his friend assumed he was more confused than normal.

Then my dad continued, "I've seen the book where all the dates of people's deaths are written down. I can find out the date of your death if you like."

"Oh no, don't worry," said his friend. "I'd rather not know!"

A couple of days later, my dad's health took a sudden turn for the worse. He was diagnosed with pneumonia and died on the Thursday, just as he had predicted.

There are many other historical cases of people predicting their own death, such as Mark Twain, who was born in 1835 as Halley's Comet was passing the Earth. Halley's Comet appears every 74–79 years, varying according to the gravitational pull of planets. A year before his death, Twain wrote, "I came in with Halley's Comet in 1835. It is coming again next year and I expect to go out with it. It will be the greatest disappointment of my life if I don't go out with Halley's Comet." Twain died of a heart attack on 21 April 1910, just as Halley's Comet was passing the Earth again. (Incidentally, as a young man Twain also had a precognitive dream of his brother Henry's death, which stunned him with its accuracy and led to a lifelong interest in parapsychology.)

Another puzzling case is that of Mikey Welsh, bass player of the band Weezer. On 26 September 2011, Welsh wrote on his Twitter account, "dreamt i died in chicago next weekend (heart attack in my sleep). need to write my will today." Later the same day, in a reply to his original tweet, he wrote, "Correction – the weekend after next." The weekend after next, while in Chicago to attend a concert, Welsh died of a heart attack in his hotel room. Although a toxicology report was inconclusive, his death was linked to a heroin overdose. (At the time of writing, the tweet is still online.[21])

The idea that the date of death is fixed is also suggested by some NDEs. Occasionally, when they meet guides or deceased relatives, people are informed that they must return to their bodies as they're not meant to die at that time. This happened to a friend of mine called John, who had an NDE during a heart transplant operation. He encountered his deceased father, who was surprised to see him and told him, "You shouldn't be here – it's not your time yet." In Chapter 5 I briefly told the story of a woman who had a "personal flashforward" in an NDE during childbirth. The guides who showed her visions of the future also informed her that: "there is a time for me to die, and that particular time when I was giving birth was not it."[22]

My Own Precognitive Experiences

If, as we saw at the beginning of this chapter, precognition is relatively common, there is a strong chance that you, the reader of this book, have had a precognitive dream at some point. And what about me, the writer of this book?

I have had several precognitive dreams that were so specific and accurate that I have no doubt that they were authentic glimpses of the future. I hope this isn't a reflection of my personality, but my precognitive experiences belong to the category of "completely trivial incidents" (although some sports fans may not agree with this categorization). For example, in September 2001, England played Germany at football, in a World Cup qualifying match. I had arranged to watch the match at my friend's house, together with his German girlfriend. The night before the match, I dreamed that I was sitting in my friend's living room watching the match, which was still in progress. The score on the TV screen read "England 4, Germany 1".

When I woke up and recalled the dream, I thought, *What? That's impossible! England could never beat Germany by that score.* Germany had been one of the best – if not *the* best – international football teams for decades, whereas England had a long history of underachievement and mediocrity. In fact, England had only beaten Germany once in 35 years, a narrow 1–0 victory a few months earlier.

But as it turned out, England outplayed Germany. After 66 minutes, they scored their fourth goal, to take a 4–1 lead. I had been thinking about my dream all the way through the match, and as the screen switched to the same score I had dreamed on the screen, I was filled with a very eerie feeling. About ten minutes later, England scored again, and the game finished 5–1. The score was 4–1 for those ten minutes or so, and it seemed logical to conclude that I had somehow "caught a glimpse" of a moment during that period. Just to get the result into perspective for those who aren't familiar with the world of soccer, it was the first time in 45 years that any team had scored five goals against Germany.

I have had two other similar dreams in which I've predicted the outcomes of sports matches. Of course, if I regularly had such dreams and most of the outcomes were wrong, the successful ones could be ascribed to chance. However, I can only recall six dreams in total in which I saw the results of sports matches. I have no idea why these three other dreams were wrong, but a "hit rate" of 50 per cent is certainly significant, when the details were specific and accurate. Trivial though the experiences were, they were certainly enough to convince me of the reality of precognition.

Incidentally, on the subject of sport, it's interesting that there is a whole subset of precognitive experiences relating to sporting events. For example, in 1946, a man called John Godley – then a student at Oxford University – dreamed he was reading a list of horse race winners in a newspaper and saw the names Bindal and Juladdin. The next morning, he checked a newspaper and found that two horses with those names were running that day. With a group of friends, he decided to risk a bet. Both horses won, netting the group a large sum of money. This happened to Godley several times over the next few years. The fourth time he had such a dream he made a written statement of his predictions (again involving two horses), which was witnessed by several people, sealed in an envelope, stamped by a post office official and locked away until the day of the race. When this prediction came true Godley became famous around the world.[23]

Empirical Research

A sceptic might argue that, although the above cases may sound impressive and some of their details have been verified, they don't constitute real scientific evidence. They are just anecdotal. For phenomenon to be accepted as reality we require empirical evidence from rigorously controlled scientific studies.

Fortunately, there is an abundance of such evidence from precognition experiments. In 1989, the researchers Charles Honorton and Diane Ferrari analysed the results of 309 precognition experiments published between 1935 and 1987, involving more than 50,000 participants and published in 113 scientific articles. They found a highly significant success rate (the odds against which were 10^{24} – ten to the power of 24 – to one), which far outweighed any possible bias due to selective reporting (that is, when researchers avoid publishing studies with negative results).[24] Recent experiments have measured physiological responses before images are presented on computer screens and found a significant correlation between the type of image and the physiological "pre-reaction". For example, participants seemed to "know" when a startling or violent image was going to appear, registering higher levels of arousal. A meta-analysis of such "presentiment" experiments between 1978 and 2010 showed an even more significant positive result than Honorton and Ferrari's analysis.[25]

In 2011, the eminent psychologist Daryl Bem – at the present time, Professor Emeritus at Cornell University – caused a stir in the scientific community when his paper "Feeling the

Future" was published in the prestigious academic *Journal of Personality and Social Psychology*. The paper described the results of nine experiments involving more than 1,000 participants, eight of which showed significant evidence for precognition. Across a variety of different procedures, Bem found that his participants seemed to be able to "intuit" information before it appeared. In a simple example, they were shown a pair of curtains on a computer screen. They were told that an image would appear behind one of the curtains, and simply had to click the correct curtain. At that point an image was randomly generated and equally likely to appear behind either of the curtains. Over hundreds of trials, there was a significantly positive hit rate. Since no image was actually there at the moment the participants made their choice, this was seen as evidence of presentiment.[26] The editors of the journal quite correctly remarked that Bem's findings "turn our traditional understanding of causality on its head".

Bem encouraged other researchers to repeat his experiments and many did so over the next few years. Perhaps even more significantly than the original experiments, a meta-analysis of 90 attempted replications of the experiments (involving 12,406 participants in 33 different laboratories) showed a highly significant positive result. Some were unsuccessful but, overall, there was a massively significant overall success rate, which exceeds the Bayes Factor rule of thumb for establishing "decisive evidence" (a standard statistical tool used by scientists) by a factor of 10 million.[27]

Other researchers have investigated the possibility of making money through precognition. In 2014, ten participants in the US attempted to predict the movement up or down of the Dow Jones Industrial Average. The predictions were correct in all seven trails, leading to a profit of $16,000.[28] In a similar 2017 study, 15 participants who had previously shown some signs of psi ability attempted to predict the movement of the German stock market. In 48 trails, 38 were correct, amounting to a highly significant rate of 79.16 per cent.[29] (The researchers decided not to use the study as an opportunity for profit, only investing tiny amounts of money, but made a significant profit in percentage terms.)

The Star Gate Programme

In my view, the above data certainly constitutes scientific proof. As Jessica Utts, President of the American Statistical Association and Professor of Statistics at the University of California, has stated, "Using the standards applied to any other area of science, it is concluded that psychic functioning [a general term covering psi abilities such as clairvoyance, telepathy and precognition] has been well-established."[30]

Utts's investigations into the results of psi experiments began in 1995, when she was commissioned by the US Congress to examine data from the Star Gate research programme, a long series of psi experiments funded by the US government. From the early 1970s, the programme was run by two physicists at Stanford Research Institute, who worked with individuals renowned for their psi abilities. In a typical early experiment, one of the researchers would visit a location and the psychic would be asked to describe or draw the scene. It soon became apparent that it was difficult to make a distinction between clairvoyance and precognition. In some cases, the psychic would accurately describe a location before the researcher arrived there. In other cases, psychics accurately described events before they occurred, such as when a psychic named Paul Smith gave a detailed account of a missile strike on an US warship that didn't occur till the following Monday. (In later experiments, psychics would accurately describe scenes based merely on geographical co-ordinates and managed to locate lost planes and provide intelligence about Soviet weapons developments.)

The Star Gate programme ran until 1995, and after spending many months analysing the data, Utts concluded that the psychics' success rate was far above chance and could not be explained in terms of methodological issues or fraud. As she stated in her Presidential Address to the American Statistical Association in 2016, "The data in support of precognition and possibly other related phenomena are quite strong statistically and would be widely accepted if they pertained to something more mundane."[31]

Small "Effect Sizes"

Even academics who are sceptical about precognition admit that the above empirical evidence is significant. However, sceptics sometimes seize on the fact that the positive results – or "effect sizes" – of precognition experiments with the general population tend to be relatively small. In layman's terms, although the results of the experiments are consistently above the level of chance, they aren't a *high* level above chance. Sceptics sometimes refer to the axiom that "extraordinary claims need extraordinary evidence."

However, this argument is unfair and illogical for several reasons. First, it's arguable whether precognition really *is* so extraordinary, as so many people have experienced it. It only seems extraordinary in the context of the belief system of scientific materialism, which insists that time is linear and human beings are purely physical beings whose thoughts are nothing more than chemical signals. Also, to argue that precognition requires a higher level of proof than other phenomena seems prejudicial, like arguing that certain criminals need a higher level of evidence to prove their innocence.

In addition, as I've pointed out above, precognition (and related abilities like telepathy) is a talent that not everyone possesses, and which only a small minority of people demonstrate at a high level. We would never expect experiments with the general population to show a high-level ability to compose music or write poetry, so why should we expect the same of psi experiments? It's no coincidence that the Star Gate programme – whose results were *far* above chance – involved gifted psychics. The same is true of the study on the German stock market (see page 172).

This links to another issue. Like other creative abilities, psi manifests itself most readily in a relaxed state, when the normal chatter of the conscious mind is quiet or absent. This is why, as we've seen, most precognitive experiences take place in dreams. To paraphrase my point in the previous paragraph, we wouldn't expect people to compose music or poetry in the clinical, pressurized environment of a laboratory, so why would we expect people them to demonstrate psi abilities in such an environment?

In my view, it's very impressive that, despite these factors, large-scale precognition experiments with the general population have had statistically significant results. Even though small, the effect sizes for psi abilities are, in the words of the psi investigator Professor Chris Roe, "broadly on a par with any other subdisciplines of psychology".[32] In a paper called "The Meaningfulness of Effect Sizes in Psychological Research", the German researchers Thomas Schäfer and Marcus Schwarz analysed a random selection of 100 empirical studies from different areas of psychology, finding an average effect size similar to psi research.[33] Across the whole of science, small-scale effects are taken as evidence. In one medical study of the effects of aspirin with over 22,000 participants, the treatment was concluded as beneficial based on a much smaller positive effect than standard psi experiments. This has also been the case with studies of medical interventions for conditions such as polio, convulsions, blood clots and AIDS.[34]

In other words, when small scale effects occur consistently in large numbers of people, scientists accept it as evidence. Surely, by that criterion, precognition has been proven.

When we add to this the highly significant results that have emerged from experiments with especially gifted psychics, together with the reports such as those we discussed earlier in this chapter, then the case becomes so strong that it is surely impossible to refute.

The Significance of Precognition

Precognition makes complete sense in terms of the vision of time we have been developing over the last few chapters. It is completely compatible with the spatial concept of time that appears in mystical experiences, near-death experiences and life reviews, where the future and past co-exist with the present. All past, present and future events are part of the same panorama, laid out before us. In precognitive experiences, we slip free of our normal linear perspective and glimpse some of the events from the panorama that we have yet to encounter in the present.

The pioneering French psychical researcher Eugène Osty worked closely with a group of psychics over many years, who described this spatial vision of time. One psychic told him that he saw time as if it was projected "on a semicircular screen", with the past to the left, the present in the middle and the future to the right. Another psychic felt that she was standing in the middle of a stream of time, able to look downstream to the past and upstream to the future.[35]

Of course, it's significant that most precognitive experiences take place in altered states of consciousness – in this case, usually in dreams. In the dream state, just as with all altered states, our normal psychological processes shift and our normal self-system may dissolve away. We may enter a transliminal state in which we gain access to the future. What differentiates gifted psychics Lorna Middleton or Alan Hencher from other people is that they have more ready access to this transliminal precognitive state. We could compare them to

THE FUTURE IS WRITTEN

great athletes such as Lionel Messi or Don Bradman, who slip easily into altered states and hence into a different timeworld.

Retrocognition

Shouldn't all this apply to the past too? If all past events are part of a panorama of spatial time, then surely we should able to re-experience the past in the same way that we can glimpse the future?

Indeed, there is an equivalent ability to precognition that focuses on the past. The early paranormal researcher (friend of William James and fellow president of the Society for Psychical Research) Frederick Myers, termed this "retrocognition". This ability is much less well known than precognition, mainly because it is much harder to substantiate both in research and personal experience. While precognition can be tested by finding out whether predicted events come to pass, there is no way to validate visions of events that have already happened. Moreover, retrocognition is difficult to distinguish from ordinary memory. For example, it could be that we occasionally have retrocognitive experience in dreams, relating to world events or past personal experiences. But if we know the event has already happened, the experience may not strike us as remarkable. No matter how vivid it is, we will probably presume it is nothing more than ordinary memory.

Nevertheless, it's worth looking briefly at some examples of retrocognition. In fact, I believe that there is one form of retrocognition that is fairly common, and which I myself experience quite regularly.

One fascinating altered state of consciousness that we haven't discussed so far is hypnagogia, the transitional state between being awake and asleep. It is well known that this state can give rise to extraordinary creativity. Many scientists and inventors have reported moments of inspiration and insight while dozing. For example, the concept of coordinate geometry occurred to René Descartes when he was half asleep in bed, while the physicist Nils Bohr had a vision of the structure of the atom while drifting off to sleep (and won the Nobel Prize as a result). It's also common for songwriters or composers to keep notebooks or recording devices on their bedside tables to capture melodies or lyrics that come to them in the hypnagogic state. The most recorded song of all time (that is, the song that has the most cover versions by other artists) came to Paul McCartney of the Beatles in his sleep. One morning, he woke up with entire melody of "Yesterday" playing in his head, and went straight to the piano to work out the notes and to find the chords to go with them. The reason for this inspired creativity is that on the verge of sleep, the boundary of the self-system becomes very soft, allowing insights and inspiration to flow through. In other words, hypnagogia is a highly transliminal state. And as such, it may allow us to access different timeworlds.

I find the hypnagogic state very pleasant and have developed the ability to hover between sleep and wakefulness for long periods. It's an opportunity to experience deeper aspects of my being, beyond my conscious mind. I feel a sense of expansiveness and harmony, as if I've connected to a source of blissful energy deep inside me. I sometimes see colourful geometric shapes, more beautiful and translucent than any normal forms. However, probably the most significant feature of my hypnagogic experiences is what I call "timeslips".

As I lie there, floating on the brink of sleep, I almost always feel a vivid connection to previous times of my life. I feel that, at the same time as being here in the present, I'm lying on the sofa in my flat in Germany 30 years ago, in my bed in Singapore 20 years ago or on the sofa in my present house 10 years ago, having a nap before going to pick up my kids from school. The sensation is much more vivid than memory. There is a clear sense of being *there* again, at these earlier times and places. I have a strong feeling of the atmosphere of the places where I lived, and of my general state and life situation at that time. I feel strongly that those past times still exist, as if I've travelled through a wormhole and emerged at an earlier point in my life. As with NDEs and the life review, it feels as if time becomes spatial and the past exists alongside the present.

Such timeslips can occur in other circumstances, too. It may not be necessary to hover between sleep and wakefulness but simply to be in a relaxed state with our minds unfocused and quiet, allowing our self-boundary to become soft. One famous example was described by the French novelist Marcel Proust, close to the beginning of À la Recherche du Temps Perdu (In Search of Lost Time). The narrator (clearly Proust himself, thinly disguised) describes how eating a soft biscuit soaked in tea becomes a revelatory experience, reconnecting him with his past. At first, he feels an overpowering sense of joy, a sense that "the vicissitudes of life had become indifferent to me, its disasters innocuous, its brevity illusory . . . I had ceased now to feel mediocre, contingent, mortal."[36]

Proust soon realizes that the source of this joy is that he has transcended linear time and returned to his childhood: "The taste was that of the little crumb of madeleine [a soft biscuit] which on Sunday mornings at Combray . . . when I went to say good day to her in her bedroom, my aunt Léonie used to give me, dipping it first in her own cup of real or of lime-flower tea."[37] Following this, Proust becomes aware of the "vast structure of recollection" held by taste and smell, as the long dormant past – of his aunt's house, the village square, the streets, gardens and the park – springs to life. Again, it's as if the past still exists, and the taste of the biscuit has opened a wormhole through which Proust has travelled back to his childhood.

Perhaps you've had a similar experience while listening to a song on the radio that you associate with an earlier phase of your life, or while revisiting a house you used to live in. Suddenly a wormhole opens and the past becomes as vivid as the present. You feel exactly as you did back then, precisely sensing the atmosphere of your life, the problems you were facing, your hopes and fears and ambitions. Of course, these experiences could be nothing more than unusually vivid memories. But since – like precognition – retrocognition makes complete sense in terms of the spatial model of time we are developing, it would be surprising if it didn't manifest itself in our lives occasionally.

Historical Retrocognition

As with precognitive glimpses of world events, retrocognition can operate beyond our normal lives. Just as some people (like Lorna Middleton or Alan Hencher) have strong precognitive abilities, there are some people who seem to be naturally susceptible to retrocognition. For example, the historian Arnold Toynbee reported many experiences when he believed he was carried down a "time pocket" to witness historical scenes. As he put it, he would be "transformed in a flash from a remote spectator into an immediate participant, as the dry bones take flesh and quicken into life".[38]

In 1921, for example, Toynbee was in the theatre of the ancient city of Ephesus in Turkey, when suddenly he found himself witnessing a scene that had taken place there almost 2,000 years ago. The empty theatre suddenly "peopled itself with a tumultuous throng as the breath came into the dead and they lived and stood up upon their feet".[39] Two Christian missionaries, Gaius and Aristarchus, had been taken to the theatre, after coming to the town to denounce pagan worship of the goddess Diana. Toynbee heard the crowd chanting "Great is Diana!" and wondered if the two men are going to be beaten to death. The town clerk intervened, trying to reason with the crowd, but at that moment the scene faded.

Of course, just as personal retrocognition can be dismissed as vivid memory, Toynbee's experiences could be ascribed simply to an active imagination. This illustrates the difficulty with investigating retrocognition, which means that it can rarely rise above the level of anecdotes.

As one final point, it's possible to explain some reported sightings of ghosts in terms of retrocognition. For example, in 1953, an 18-year-old apprentice heating engineer named Harry Martindale was in the basement of a large house in York, England, replacing a boiler. He heard a series of trumpet calls that grew steadily louder. Looking up, he saw a Roman soldier emerge through the wall, followed by the trumpet player, a horse and a group of around 20 soldiers, walking in pairs. They were wearing green tunics, carrying round shields and short swords, with sandals laced up to their knees. For some reason Martindale could only see them from their knees upward. Terrified by the apparition, he took two weeks off work to recover.

This case became well-known because some of the details Martindale mentioned were later verified. Initially, when his experience was reported, it was dismissed because at that time there was no evidence that Roman soldiers wore green or used round shields. However, later excavations of the basement of the house in York found that a Roman road did run through it. The road was about 38 centimetres (15in) below the cellar floor, which would explain why Martindale couldn't see below the soldiers' knees. Moreover, further research found that auxiliary Roman troops in the North of England carried distinctive round shields and wore green tunics and sandals laced up to their knees.

There are certain aspects of this case that differentiate it from conventional ghost experiences. For example, Martindale stated that the soldiers appeared very clear and solid, and seemed completely unaware of his presence, making no attempt to communicate. It seems more likely that, rather than seeing the ghosts of deceased soldiers, he passed back 1,500 years through a timeslip.

Implications

It is impossible to seriously study precognition (and to an extent, retrocognition) without concluding that our normal experience of time is illusory. As the French investigator Eugène Osty commented, "Time, as well as Space, is penetrable by the faculty of super-normal cognition, just as if Time were but an illusory creation of the human mind."[40] After studying cases of precognition collected by the Society for Psychical Research, the early British researcher Herbert Saltmarsh stated "our ordinary idea of the nature of time is clearly inaccurate . . . we are bound to admit that the future does exist in some sense now – at the present moment."[41]

Indeed, the fact that precognition has such massive implications is probably one of the reasons why some people are so keen to dismiss the evidence for it. As with NDEs and the life review, it's more convenient to reject anomalous phenomena rather than entertain the idea that our common-sense view of reality may be limited or even false. However, it should be clear by now – not just from precognition but from all the anomalous time experiences we've discussed in this book – that the latter is the case.

What needs to be questioned is not the evidence for NDEs or precognition but our normal view of linear time. As another precognition researcher, Robert Thouless, puts it, "If the facts of precognition are in conflict with our customary ways of thinking about time, then our ways of thinking about time need changing."[42]

In the following chapter, we will broaden our discussion by examining further evidence that our normal perception of time is illusory. One argument that sceptics sometimes make is that precognition cannot be real because it "breaks the laws of science". In the next chapter, we will see that this is not at all the case. In fact, the spatial vision of time that is implied by precognition (and by near-death experiences and the life review) is supported by many theories from modern physics.

CHAPTER 9

BEYOND LINEAR TIME
Perspectives from Physics

Unlike most modern-day scientists, Isaac Newton was an extremely religious man. He devoted much of his life to biblical studies, writing works such as *Observations upon the Prophecies of Daniel and the Apocalypse of St John*. At the same time as uncovering the fundamental laws of the physical world, Newton used his acute powers of reason to try to unravel – with rather less success – theological mysteries such as the date of Christ's Second Coming and the Apocalypse.

Nowadays it's common to see science and religion as opposing forces, but Newton saw his scientific endeavours as an attempt to understand and explain God's creation. As he wrote in his primary scientific work, the *Principia*: "This most beautiful system of the sun, planets and comets could only proceed from the counsel and dominion of an intelligent and powerful Being."[1] Newton was aware that his laws of motion and gravitation could imply that the universe was nothing more than a great machine. However, he believed – as many advocates of "intelligent design" still do – that the perfect order and balance of the universe was only explicable in terms of a divine power. As he wrote to the theologian and scientist Richard Bentley in a letter of 1692, "gravity may put the planets into motion but without the divine power it could never put them into such a Circulating motion as they have about the Sun, & therefore for this as well as other reasons I am compelled to ascribe the frame of this Systeme to an intelligent agent."[2]

Time was an integral part of Newton's mechanistic vision. He saw time as "absolute, true and mathematical . . . [It] flows equably without regard to anything external."[3] Time was a fundamental feature of the universe, like space, a background on which all events took place. It flowed at the same pace everywhere in the universe, whether anyone was there to experience it or not. The entire universe moved in unison, from the past to the present to the future, like a giant machine locked in motion.

Newton's mechanistic vision dominated physics for more than two centuries – although sadly for him, later scientists felt they could account for the clockwork-like precision of the universe without a creator. Darwin's theory of evolution showed that complex life forms developed through natural processes over aeons of time, obviating the need for the handiwork of God. Other developments, such as the discoveries of fossils from millions of years ago, damaged the authority of the Bible that Newton studied so intensively.

Einstein's Relativity

At the beginning of the 20th century, many scientists believed that their understanding of the world, founded on Newton's laws, was basically complete. As the American physicist Albert A. Michelson wrote in 1903, "The more important fundamental laws and facts of physical science have all been discovered, and these are now so firmly established that the possibility of their ever being supplanted in consequence of new discoveries is exceedingly remote."[4]

This turned out to be one of the most premature and hubristic scientific statements ever made. Just two years later, Albert Einstein published a series of papers that revolutionized physics, upending the Newtonian vision of the universe. In his theory of relativity, Einstein showed that time was not absolute and did not flow "equably". Time did not exist prior to material objects and could not exist without them. It was not the backdrop on which events unfold but a flexible local phenomenon.

Einstein showed that time depends on both gravity and speed. Time runs more slowly as gravity increases. In his 1915 paper on his Theory of General Relativity, Einstein described how massive bodies with a strong gravitational pull slow down time in their vicinity. When massive stars collapse in on themselves and form black holes, their gravitational pull is so great that light can't escape. If you could stand on the edge of a black hole, it would seem as if time had stopped. Any objects falling toward the hole would appear frozen.

On a slightly more local level, time runs more slowly on the surface of Jupiter than on the Earth, because of its stronger gravitational field. And even on the Earth itself there is some variation. Time runs slightly more slowly at sea level than on mountains, because there is more gravity near the surface of the Earth. Even more microcosmically, if you live to be 80, your head will have aged 300 nanoseconds more than your feet!

In addition, time runs more slowly as speed increases. If you were on a spaceship travelling at 87 per cent of the speed of light, time would move twice as slowly relative to an observer on Earth. If you kept increasing your speed, time would slow further still, at least until you neared the speed of light. According to Einstein, it is impossible to reach the speed of light, which would require an infinite amount of energy. (Although some physicists have suggested that hypothetical particles called tachyons may exist, which have no mass and can exceed the speed of light.)

Einstein's theory also showed that the sequence of events is relative to observers at different vantage points. Whereas a person on the ground might witness two events happening concurrently, a person travelling at high speed above them might register a time lag. In another context, a person travelling at high speed might perceive two events in reverse order to a person on the ground.

The Block Universe

Following Einstein's initial papers, one of his tutors, the mathematician Hermann Minkowski, became aware of an important implication of relativity: that time could not be considered in separation to space. Newton had only described three spatial dimensions – height, length and width – with time as a backdrop. But Einstein's theory meant that time could not be set apart. Minkowski developed the notion of "space–time", uniting time with the three dimensions of space to create a four-dimensional "block universe".

This model is still accepted by the majority of physicists. Time doesn't flow, it is simply static, spread out like a panorama. As the eminent British physicist Roger Penrose has stated, "The way in which time is treated in physics is not essentially different from the way in which space is treated . . . We just have a static-looking fixed 'spacetime' in which the events of our universe are laid out!"[5] From our perspective, it may seem as if time is flowing, but this is just an illusion of perspective. Here inside the block universe, the point of *now* is equivalent to *here*. Just as there are an infinite number of locations in space, all equally real at this moment, there are an infinite number of points in time, all equally real and present. In other words, the past, present and future co-exist concurrently. Physical objects –such as human bodies – do not move through time but create "worldlines" across space–time.

Many modern physicists have gone beyond the four-dimensional model of Einstein and Minkowski. There are varieties of string theory that suggest the existence of many dimensions. Superstring theory suggests 10 dimensions, M-theory suggests 11, while Bosonic superstring theory suggests 26 dimensions. Beyond this, there are theories of a multiverse, or the potential infinite parallel universes of the "many worlds interpretation" of quantum physics. Obviously, these models imply notions of time that are very different to our normal experience. For example, the contemporary British cosmologist Bernard Carr has developed a multi-dimensional model that incorporates several different dimensions of time. According to Carr, there is a hierarchy of times, linked to a hierarchy of different states of consciousness. Such multi-dimensional models can explain the passage of time, even in a static block universe.[6]

Eternalism and Presentism

In philosophy, the idea that the past and future are as real as the present is known as "eternalism". This contrasts with "presentism", which insists that only the present is real. For eternalists, time is like a lake stretching out wide in all directions, static and unflowing. For presentists, time is a river flowing ceaselessly in one direction as each moment arises and dissolves.

Of course, presentism describes our normal experience of time. The idea that the past and future are as real as the present seems to outrage common sense. However, there is very little support for presentism in physics. As we have just seen, eternalism is supported by the block universe theory. But even elsewhere in physics, the concept of "now" – a special moment of the present that separates the past from the future – has surprisingly little traction. As the philosopher Craig Callender has remarked, the equations of physics are "like a map without the 'you are here' symbol. The present moment does not exist in them, and therefore neither does the flow of time."[7]

One normal aspect of our experience is that time flows in one direction – forward, away from the past and into the future. In everyday life, all events in the present have a future effect. I drink a boiling cup of coffee and burn my tongue. Rain falls from the sky and my clothes get wet. This is also the basis of all the intentional acts that fill our days. We buy food so that we can cook later. We exercise for the sake of our future health. We save money for our retirement.

However, this common-sense "unidirectional" perception is largely absent from physics. There is just one basic law of physics that implies a forward direction of time: entropy, the law of decay. Over time, all systems inevitably lose energy and order, deteriorating toward chaos. Entropy is always lower in the past, and higher in the future. But entropy may not be a fundamental feature of the universe. It may only exist at a macrocosmic level, affecting large collections of particles. According to the Italian physicist Carlo Rovelli, author of *The Order of Time*, entropy is not part of the "elementary grammar" of the world.[8]

Besides entropy, it makes no difference to the laws of physics whether time is conceived of as flowing forward or backward. This applies not only to Einstein's laws, but also to the equations of Maxwell (which describe electricity and magnetism), those of Schrödinger (which describe the microcosmic world of quantum physics) and even Newton's laws. These equations and laws are time symmetric and time reversible, in that they work just as well in either direction of time. Similarly, in particle physics, space–time diagrams can be interpreted as showing particles moving either forward or backward in time. The interpretations are mathematically identical and therefore equally valid.

The Quantum Revolution

Another way in which Albert Michelson's statement was wildly premature is that he failed to anticipate the advent of modern quantum physics. In the 1920s, the discoveries and theories of quantum physicists such as Nils Bohr, Werner Heisenberg and Ervin Schrödinger further eroded the solid certainties of the Newtonian world. Just as Einstein and Minkowski uncovered the strange variability of the macrocosmic universe of planets and stars, quantum physicists showed that the microcosmic world of atomic and sub-atomic particles was uncanny and unpredictable. In the words of the contemporary physicist Paul Davies and the science writer John Gribbin, with the advent of quantum physics, "Newton's deterministic machine was replaced by a shadowy and paradoxical conjunction of waves and particles governed by the laws of chance, rather than the rigid rules of causality . . . [S]olid matter dissolves away, to be replaced by weird excitations and vibrations of invisible field energy."[9]

Quantum physics has shown that the laws of Newtonian physics do not apply to particles smaller than the atom. In fact, they are flatly contradicted. Particles behave in ways that make no sense in Newtonian terms. They may vanish and reappear seemingly at random. They can affect each other's movements even if they're miles apart, as if the normal rules of causation – and time and distance – no longer apply. They can lose their form as particles and appear to behave as waves instead, as if they are manifestations of a more fundamental reality that can express itself in different forms. Particles even have the ghost-like ability to pass through seemingly solid barriers (which is referred to as "quantum tunnelling").

The subatomic world is one of connection rather than separation. Particles can instantaneously connect with each other across large distances, without any signal passing between them and without any decay in the strength of the connection. In 2015, a group of American physicists found that photons (particles of light) 200 metres (656ft) apart displayed interconnected behaviour, as if they were communicating across the distance. Even more remarkably, in 2017 a group of Chinese physicists showed that such "entanglement" between particles could occur at a vast distance of 1,400 kilometres (870 miles), using a "quantum satellite" floating high above the Earth.[10] Of course, this makes absolutely no sense in terms of Newtonian physics.

Equally, quantum physics reveals a connection between the observer and the world they observe. The world doesn't seem to be "out there" as an independent, objective reality, but is co-created by our consciousness. As the British physicist John Wheeler wrote "Nothing is more important about the quantum principles than this, that it destroys the concept of the world as 'sitting out there', with the observer safely separated from it . . . In some strange way the universe is a participatory universe."[11] The behaviour of particles – including whether they appear in the form of waves – is affected by human observation, as if human consciousness is inextricably linked to the physical world.

This quantum strangeness also manifests itself in relation to time. One of the most well-established concepts of quantum physics is "superposition", which means that the location of particles is indeterminate. Until they are observed or measured, it is impossible to know where they are. Contrary to common sense, they are not anywhere in particular. In recent years, many quantum physicists have come to believe that this is also true of time. The difference between the past, present and future is flexible, so that sequence and causal relationship are undetermined. As Carlo Rovelli puts it, "Just as a particle may be diffused in space so, too, the differences between past and future may fluctuate: an event may be both before and after another one."[12]

This "superposition of sequence" was first suggested in a thought experiment by the Austrian physicist Caslav Brukner in 2012.[13] Three years later, the theory was verified experimentally when a team of physicists at the University of Vienna led by Philip Walther observed a photon (a particle of light) pass through two gates (A and B) in indefinite sequence. It was impossible to say whether the photon passed through A before B, or B before A. Its movement was a superposition of both sequences.[14]

Retrocausality

In quantum physics, the indeterminate nature of time is also highlighted by the concept of "retrocausality". In the common-sense Newtonian world, cause and effect always have a future orientation. It's clearly impossible for events in the present to affect the past or for future events to affect the present. But in the quantum world, this is not necessarily the case. Many quantum physicists believe it is possible for causal influences to travel back in time from the future to the present and from the present to the past. This is partly a way of making sense of the "time-symmetry" of the laws and equations of physics.

Quantum physics suggests that physical events occur when quantum waves "collapse" through observation or measurement. Until that point, all events just exist in the form of probabilities. When a new physical state is created in this way, probabilities of future events are also generated. So far, so logical. However, the collapse process also affects past events. According to the American mathematician and physicist Henry Stapp, the past processes related to the probabilities that were eliminated disappear, leaving no trace. The records of the processes that led up to the event only include the elements related to what actually did happen. In this way, even the past is created by the act of observation.[15] As Stephen Hawking and Leonard Mlodinow put it, "We create history by our observation, rather than history creating us."[16] This "backward in time" effect may sound esoteric, but it is the basis of many phenomena studied by quantum physicists. This includes the famous "delayed choice" experiment devised by the British physicist John Wheeler, which showed that whether a photon behaves as a wave or a particle depends on future measurements.

Another retrocausal quantum theory is the "transactional" interpretation of quantum mechanics, developed by the American physicist John Cramer. According to him, quantum events are an interaction between waves that move both forward and backward in time. In the metaphor used by Cramer, they are a "handshake" across space–time. Waves from the past are "offered" and then confirmed by waves from the future. Time becomes a two-way street, in which the future determines the past, as well as vice-versa.[17]

Implications

What should we make of these theories and findings? What relevance do they have for the ideas and topics I've been discussing over the last eight chapters, such as time expansion and cessation experiences, near-death experiences and precognition?

On one level, Einstein's findings clearly parallel the flexible model of time we've been developing throughout this book. The universal relativity of time fits with the psychological relativity of time. There is no absolute, objective universal time, just as there is no absolute or objective psychological time. In both inner and outer space, time is elastic. In either inner and outer space, time varies according to different states – either states of mind, or states of speed or gravitation.

This even applies to the cessation of time. In both inner and outer space, time can stop in certain circumstances. In inner space, time stops when our self-boundary dissolves. In outer space, time stops in black holes, and perhaps also if an object reaches the speed of light. (As we have seen, Einstein maintained it was impossible to reach or bypass the speed of light, but some physicists disagree.)

These are only parallels, though. I'm not implying that the two types of relativity are equivalent or drawing any causal relationship between them. It is tantalizing to think they could be related, and perhaps this would fit with the notion from quantum physics that human consciousness is inextricably linked with the physical world. But there is no real equivalence between the slowing of time that occurs in intense altered states of consciousness and the slowing that occurs in celestial bodies with strong gravitational fields.

However, perhaps the block universe theory does have direct relevance for the model of time we have been developing. The block universe theory describes essentially the same spatial vision of time that I have discussed over the previous four chapters in relation to NDEs, life reviews, precognition and retrocognition. If it is possible to experience the future and the past, then they must both already and still exist, as part of a spatial landscape of time.

In other words, both the block universe theory and the material we have discussed suggest an eternalist view of time. The block universe theory fully supports the notion that linear time is a construct of the human mind, rather than a fundamental reality of the universe. This doesn't mean that the block universe theory *accounts for* the life review, or precognition or retrocognition. But it certainly supports – and allows for – such experiences.

As for quantum physics, it's difficult to draw any firm conclusions. The strangeness and uncertainty of the sub-atomic world is a graphic illustration of the limitations of the human mind. Some philosophers and scientists believe it's possible for human beings to completely understand the world, that one day we will possess theories and laws that fully explain life, the universe and everything. But we're only animals, after all. Every animal has a limited awareness and a limited intelligence, and therefore a limited capacity for understanding. Perhaps in a few million years, there will be life forms with more awareness and intelligence than us, who will be able to make some sense of quantum physics. As the American physicist Richard Feynman is reputed to have said, "If you think you understand quantum mechanics, you don't understand quantum mechanics."

However, perhaps we can draw some tentative conclusions. The concept of retrocausality suggests that the common-sense "arrow of time" is a misconception – unless we accept that the arrow can move backward as well as forward, and that the arrow can sometimes fly before the archer has shot it. The indeterminate nature of time also suggests that time emerges through observation. The act of observation confirms sequence and causality. Up to that point, linear time does not exist.

Perhaps most importantly, quantum physics suggests that time is not a fundamental property of the universe. Time seems to emerge when human consciousness interacts with an essentially timeless reality. As Carlo Rovelli has stated "Time emerges from a world without time."[18] In recent years, this view has become more and more popular among physicists, partly as a way of reconciling relativity with quantum theory. There are some aspects of the two theories that seem incompatible – for example, the warping of space–time by large celestial bodies, which makes no sense in terms of quantum theory. But if both theories are descriptions of something more fundamental, their incompatibility is less problematic. The concept of "something more fundamental" could also account for entanglement, when particles appear to connect across distance without any information flowing between them. Perhaps this only appears impossible at our emergent level. At the most fundamental level of reality, space may not exist. Distance may just be a construct that arises at the physical level – and the same applies to time.

In summary then, all these findings support the notion that time is a construct rather than an objective external reality. Modern physics points to the same conclusion that we have drawn from spiritual experiences, NDEs, the life review and precognition: that our common-sense view of linear time is limited and misleading.

Physics and Precognition

One intriguing aspect of the block universe theory is that it implies the possibility of time travel. If all the past and future events of the universe are laid out in stasis, it might be possible to journey to different points of time, just as it's possible to travel to different locations in space. This might be possible through "wormholes", for example, which – in theory at least, as they have never been detected in reality – arise when the fabric of space–time is torn open by massive bodies like black holes, creating a tunnel between different times and places.

However, there is another form of time travel that isn't just a theory, and often occurs: the mental time travel of precognition and retrocognition. From the presentist perspective, any form of time travel is impossible, as the past no longer exists and the future doesn't yet exist. But as we have seen, from the eternalist perspective, precognition and retrocogntion do make sense.

In view of this, it's surprising that some scientists reject the possibility of precognition (and other types of extra-sensory perception such as telepathy and clairvoyance) on the grounds that they break the laws of physics. Some scientists refuse to even consider the evidence for psi on the grounds that it is theoretically impossible. For example, in 2018 the psychologist Etzel Cardeña published a paper in *American Psychologist*, in which he carefully and systemically reviewed the evidence for psi phenomena and concluded that the evidence for psi is "comparable to that for established phenomena in psychology and other disciplines".[19] The following year, the journal published a rebuttal by Arthur Reber and James Alcock. The article was rather bizarre, in that the authors didn't actually engage with the evidence presented by Cardeña. They simply decided that since psi contravened the laws of physics, the evidence could not possibly be valid and so wasn't worth examining.

As they wrote "Claims made by parapsychologists cannot be true . . . Hence, data that suggest that they can are necessarily flawed and result from weak methodology or improper data analyses."[20]

It's certainly true that psi phenomena contravene the principles of Newtonian physics, but then relativity and quantum theory don't fit with Newton's laws either. In turn, psi phenomena do not contravene relativity and quantum theory. On the contrary, the block universe theory and the weirdness of the subatomic world certainly *allow for* precognition, and other psi phenomena. How can we rule out telepathy when, at the most fundamental level of reality, particles are in multiple places at the same time and connect instantaneously across great distances without any signal passing between them? How can we rule out precognition when, at the most fundamental level of reality, time does not separate itself into different tenses and quantum events can reach backward in time, so that the future influences the present and the present influences the past?

I'm not suggesting that quantum theory can *explain* telepathy and precognition. But when reality is so strange, it seems foolish to outlaw such psi phenomena. As the cosmologist Bernard Carr has stated, there is plenty of "space for psi" in contemporary physics.[21] (In Carr's own model, some paranormal phenomena are interpreted as influences or "intrusions" from higher dimensions.[22]) Or as Jessica Utts, whose analysis of the evidence for precognition and other psi phenomena we referred to in the last chapter, has pointed out, "Physicists are currently grappling with the concept of time and cannot rule out precognition as being consistent with current understanding."[23]

At both the ultimate microcosmic and macrocosmic levels – from sub-atomic particles to vast celestial bodies – it is our common-sense view of time that merits scepticism. Modern physics shows that the essential reality of time appears to be vastly different to our common-sense conception. And as this book has illustrated, time expansion and time cessation experiences point to the same conclusion.

Some physicists – particularly quantum physicists – separate their work from their worldview and don't consider the implications of their theories and findings. For example, I once chatted to a university colleague who was doing research on quantum theory and asked him how he dealt with the weirdness of the quantum world. He recoiled in horror and remarked, "I just don't go there! I just do the work and don't worry about the rest. I just hope that eventually a theory will come along that makes sense of it all."

However, Einstein felt obliged to acknowledge the implications of his theories. He allowed the strangeness of theoretical physics to inform his perspective on life. One example of this was his curiosity about telepathy. He accepted the possibility of its existence, and even wrote a foreword to a book on the topic, *Mental Radio* by Upton Sinclair, published in 1930.

In a more general way, Einstein's famous good humour and levity (as seen in the famous photo of him as an old man sticking out his tongue) were probably the result of his awareness of the fleeting, insubstantial nature of time, and hence of human life. On the 15th of March 1955, his close friend and collaborator Michele Besso passed away. At the time, Einstein himself was seriously ill, and aware that he was close to death himself (in fact, he was to die just a month or so later). Einstein wrote a letter of consolation to Besso's family, but it's tempting to think that he was also consoling himself in the face of impending death. As he wrote, "He has preceded me briefly in bidding farewell to this strange world. This signifies nothing. For us believing physicists the distinction between past, present and future is only an illusion, if a stubborn one."[24]

CHAPTER 10
TRANSCENDING TIME

Softening the Self-Boundary

The late 18th-century German philosopher Immanuel Kant lived very firmly in a world of time. He was such a stickler for routine that his fellow citizens of Königsberg (now the Polish city of Kaliningrad) could set their watches by his afternoon walk. According to legend, he only failed to appear for his walk once, when he was distracted by reading a book by Jean-Jacques Rousseau. Throughout the whole of his long life (he died in 1804 at the age of 79), Kant never travelled more than ten miles from his home city, despite being an avid reader of travel books.

This means that – in an *experiential* sense – Kant's life must have been fairly short, despite its long span. As we saw in Chapter 1, routine and repetition make time pass quickly, while new experiences and environments slow down time, because they allow for increased information processing. Living according to a strict routine and never travelling is certain to accelerate the passage of time.

However, in his philosophical work Kant was sceptical about time. He had the same basic insight that many people gain in near-death experiences: that time is a mental construct. Time and space do not exist as fundamental qualities of the world. Out there beyond our minds, in reality itself, there is no time. It is simply a "category" of our minds that helps us to perceive objects and order our experience. In Kant's terminology, we "impose" time and space on the world, via our "apparatus of perception".

Kant called the fundamental reality of the world the "noumenon." As human beings, we live in the phenomenal world, the content of which is determined by our mental constructs. We cannot gain access to "things as they really are", because we cannot drop our apparatus of perception. Space and time are like irremovable spectacles through which we view the world. We can only see objects through the lens of time and space and therefore can never view the world in itself.

Beyond the Phenomenal World

Kant's insights mirror one of the essential ideas of this book: that time is not an objective reality but is created by – and varies according to – our psychological processes and structures. However, this book has also shown that in an important sense Kant was wrong. We are not helplessly imprisoned by our mental constructs. It *is* possible to take off the spectacles and step outside time. Throughout this book, we have examined what happens when the psychological processes that produce our normal time perception are altered. In other words, we have examined what happens when we remove our normal spectacles (or at least when we *partially* remove them, since we may become free of time to a greater or lesser degree, depending on the intensity of an altered state of consciousness).

Kant's error is that he wasn't aware of the significance of altered states of consciousness. He didn't realize that, in William James's words, our normal state is "but one special type of consciousness, whilst all about it, parted from it by the flimsiest of screens, there lie potential forms of consciousness entirely different."[1] Kant assumed that it is impossible to escape our normal consciousness and our normal mental constructs, when in fact we all experience occasional moments of freedom from them. As we have seen, this can happen in accidents, the Zone experiences of athletes, awakening (or mystical) experiences, near-death experiences, and so on.

Perhaps if he had lived a little longer and read the first European translations of Indian texts such as the *Upanishads* and the Buddhist *Dhammapada* – which became available during the first half of the 19th century – then Kant's perspective would have been different. This is one essential difference between Western and Eastern philosophical traditions. The basic approach of Western philosophy has been to examine the world as it appears to our normal consciousness (and more recently, to examine consciousness itself), usually on the assumption that normal consciousness is reliable and objective. On the contrary, the basic approach of Eastern philosophies such as Buddhism and Hindu Vedanta has been to *transcend* our normal consciousness in order to attain a *more* objective and reliable vision of reality.

In the last chapter of this book, we will follow a similar approach to Eastern traditions and investigate the possibility of transcending our normal consciousness, including our normal perception of time. In Kant's terminology, we will investigate the possibility of freeing ourselves from the phenomenal world, or taking off our spectacles. If our normal time perception is a flexible construct, is it possible to free ourselves from it? Can we consciously alter our experience of time – perhaps even induce time expansion experiences?

A Summary

Before we begin this closing discussion, let me summarize the basic argument of this book.

Our experience of time is not objective but is constructed and mediated by psychological processes and structures. In normal (and some mildly altered) states of consciousness, time varies mildly according to factors such as information processing and mood. But when we enter *intense* altered states of consciousness – through accidents, drugs, meditation or near-death experiences – then our experience of time alters more radically, usually by slowing down drastically.

In the most extreme altered states, such as high intensity awakening experiences and near-death experiences, we may enter a world of "no time" or "all time" – that is, a world in which time simply doesn't exist, or in which it appears to be spatial, a panorama that includes the past and future.

The aspect of our normal "self-system" most responsible for our normal experience of time is its strong boundary, which creates our sense of being enclosed inside our mental space, with the rest of the world "out there" on the other side. When this boundary fades or disappears, in certain altered states, then time itself fades and disappears. The softer the boundary becomes – as an altered state becomes more intense – the more expansive time becomes. And when the boundary dissolves completely, time disappears completely.

As the self-boundary fades, our spatial awareness also changes. We may feel that we are both inside and outside our bodies. We may feel that our identity extends into the world and other beings, bringing a sense of connection or oneness. As Kant intuited, space and time are related constructs that develop and disappear together. (This parallels Minkowksi's realization that space and time are interdependent, hence his concept of space–time.)

We are left with two conclusions, which both appear to be true. At one level – a more fundamental one that our normal experience of reality – time is panoramic. As unusual states of consciousness such as intense awakening experiences, NDEs and the life review suggest, in some sense the past still exists and the future already exists. All the past and future events of our lives exist alongside the present. In physics, the block universe theory supports this perspective.

On another level, time is not a fundamental feature of the universe at all. As many theories and findings from quantum physics suggest, time is an emergent phenomenon rather than an essential quality. At the most fundamental level, the universe appears to be timeless.

Free Will?

If the future already exists, everything in our lives is pre-determined. It doesn't matter what decisions or plans we make – we are inexorably pulled toward future events that have already happened or are already happening. In other words, we don't possess free will. We can't make a conscious intention to do anything. As we live our lives, it may seem as if we make conscious decisions, but we're simply following the pattern of pre-established events.

Free will is a serious issue for eternalism, and for precognition. If we can glimpse the future, our course in life is already set. How can we have free will when every choice we make is tied to certain actions and outcomes, which must take place because they have *already* taken place?

Many contemporary philosophers and scientists don't believe in free will. They hold that all our decisions and actions are determined by factors beyond our conscious control, such as our genes, brain activity and environmental conditioning. They believe that human beings are simply biological machines, with an illusory sense of self, and so cannot act as free agents. According to Sam Harris, for example, human intention emerges from "background causes of which we are unaware and over which we exert no conscious control". As a result, you are effectively a "biochemical puppet."[2]

However, I do believe in free will, at least to an extent. In my view, our feeling of freely making decisions is so fundamental that it can't be disregarded entirely. The clearest way that this expresses itself is when we decide *not* to do things – when we use our free will to override impulses or habitual or expected responses. For example, after deciding to give up caffeine, you might find yourself in a café, feeling an urge to order a cappuccino. But then you might use your willpower to override the impulse. You might be expected to attend a meeting or do a presentation at work and decide, out of nervousness or sheer bloody-mindedness, to go home instead . . . Then you consider the consequences of losing your job, change your mind, turn around and walk back to the office.

There are all kinds of forces that determine our decisions and actions. Free will is one of them, along with genes, brain activity and environmental conditioning. One of the main developmental tasks of our lives is to increase our free will, so that we become less dominated by other forces. In traditions like Buddhism and Yoga (and also the Christian monastic tradition), the ability to control one's behaviour, instincts and desires is an important initial stage of spiritual development.[3]

But how I can square a belief in free will with an eternalist view of time?

The best way to look at this is to see future events as *probable* rather than determined. The future events that are glimpsed in precognition – or which are laid out statically in a block universe – may not be completely fixed but include some possible variation. This makes sense from the perspective of quantum physics, in relation to superposition and the indeterminate nature of time. Perhaps in the same way that particles are never in one particular place until they are observed, we can never describe the precise nature of future events until they manifest in the present.

You might argue that this is just an argument of convenience, as it allows eternalism and free will to co-exist (and also permits inaccuracy in precognitive experiences). But precognitive experiences themselves offer some evidence for this argument. Some experiences seem to function as forewarnings of accidents or dangers. Shortly afterwards, people recognize the situation from their dreams, and sometimes manage to take preventative action.

For example, in a case from 1888 collected by the Society for Psychical Research, a woman described how she planned to travel by train to Roehampton, on the outskirts of London, to visit her sister. The night before the visit, while drifting off to sleep, she had a vision of the carriage that met her at the station overturning close to her sister's house. The vision jolted her fully awake and recurred when she started to fall asleep again. The next day she stepped into a carriage at the train station and in her words:

> Everything went on smoothly till we were driving up the lane to my sister's house, when the horse became very restive, the groom got down, but could find nothing wrong, so we went on; this happened a second and a third time, but when he was examining the horse for the third time my vision of the night before suddenly came back to me, and I told the groom I would get out and walk to the house; he tried to dissuade me, but I felt nervous and insisted upon walking, so he drove off by himself, and had only got a very short distance from me when the horse became quite unmanageable. I hurried on some men in the road to help him, but before they reached him the carriage, horse and groom were all in a confused broken heap in the hedge, just as I had seen it the night before, though not exactly in the same spot. The groom managed to extricate himself, but when I got up to him he said he was so thankful I insisted upon getting out, for he could not possibly have saved me from a dreadful accident.
>
> I had no fear of horses. I should certainly not have left the carriage but for the forewarning of the previous night.[4]

Clearly, the fact that the woman was able to take preventative action means that her dream wasn't completely accurate. It's also interesting that the woman reported the wreckage of the carriage was "not exactly in the same spot" as in her dream, which suggests some variability in the manifestation of future events.

In an example from the American psi researcher J B Rhine, a man described how, when working as an investigator for an insurance company, he was woken up early one morning by a policeman friend who asked to borrow his gun. In return, the policeman left his own gun, advising the man not to use it, and telling him that he would explain why later. That same day the policeman was called to a robbery, which turned into a shoot-out. He shot two of the robbers dead and received a minor chest wound himself. In hospital, the policeman told his friend that "I dreamed I was in a gun fight and the gun failed on the third shot. The dream was so real that I just knew I had to have a good gun before I went on duty." The insurance investigator took the policeman's gun to a firing range, and sure enough, on the third shot, the "main spring let go, rendering the gun useless".[5]

These cases clearly allow for the possibility of free will. If precognition was simply a glimpse of fixed future events, then these people would have been unable to take preventative action.

Some life previews also imply a future that is flexible rather than determined. While some future events do appear to be as real and definite as the present, others are described as alterable or provisional, dependent on the choices that a person makes. For example, in Chapter 6, we discussed the childhood NDE of a man called Tracy, who saw a detailed vision of his future but was also aware that the vision was not "set in stone". Paradoxically, it appeared both predetermined and flexible. Anita Moorjani also recalled that during her NDE (which we looked at briefly in Chapter 5), "What I felt was that all possibilities exist simultaneously – it just depends which one you choose . . . like being in an elevator, where all the floors of a building exist, but you can choose which floor to get off on."[6]

Similarly, in some NDEs people feel they have a choice whether to return to their bodies or not, and are informed of certain events that will occur if they do return. For example, as we saw at the beginning of Chapter 8, Elizabeth Krohn was told by a guide that if she returned to her body, she would have a third child – a daughter – but that her marriage would not survive. Both these events came to pass. Clearly though, the events were conditional rather than definite. If she hadn't decided to return to her body, they obviously wouldn't have occurred.

In other words, eternalism doesn't necessarily negate free will. Future events are probable rather than fixed.

Controlling Time

So, after concluding that we possess the free will to do this, let's discuss the question of whether it's possible for us to consciously control our perception of time.

In fact, there are some straightforward ways in which we often do – if only unconsciously – control our experience of time. Here we can return to the "laws of psychological time" from Chapter 1. (Hopefully that doesn't seem so long ago now, if this book has managed to keep you in a state of absorption.) Both the second and third laws imply some simple strategies of controlling our experience of time.

If you recall, the second law of psychological time is: "Time seems to go slowly when we're exposed to new environments and experiences (or inversely, time goes quickly when we're in familiar environments and have familiar experiences)." This logically implies that we can slow down time by exposing ourselves to as much new experience as possible. We can expand time by travelling to new places, facing new challenges, meeting new people, absorbing new information, learning new hobbies and skills, and so on. This stretches time by increasing the amount of information that our minds process. In my view, this is one reason why human beings love to travel and to take up new challenges and hobbies. Unconsciously, at least, we're trying to slow down time. Two weeks of travel in a strange foreign country stretches time several times longer than a fortnight in our familiar home surroundings. A day away at a workshop meeting new people and absorbing new information lasts a lot longer than a normal day at work.

We can extend this over a whole lifetime too. Two people who die at the same age can experience vastly different amounts of time in their lives. Imagine a person who spent their life in the same town, repeating the same experiences in the same routine. They spent their working life in their one and only profession and went on vacation to resorts that resembled their home environment. Then consider another person – perhaps their twin, for ease of comparison – who lived in several different countries, worked in a variety of professions and spent their vacations on adventure trips in obscure parts of the world. Also imagine that the second twin had a curious open mind, followed many hobbies, learned a lot of different skills and investigated a whole spectrum of ideas and approaches. In experiential terms, the second twin's life span would be much longer than their sibling's.

In other words, we can expand any period of time – a weekend, a two-week vacation or a whole life span – by exposing ourselves to unfamiliarity. There is an alternative approach though. Rather than changing our surroundings, we can change *the way we perceive* our surroundings. After all, there is no essential difference between our home environment and a strange foreign country. A person from a strange foreign country would find your home environment just as invigorating as you do theirs. The only difference is our *perception* of our home environment. We have *allowed* it to become familiar and dull. We have switched off to its suchness. So perhaps we can alter our perception in such a way that a familiar environment no longer appears mundane

– perhaps even as interesting and beautiful as a strange foreign environment.

One way to do this is through mindfulness, which entails giving our whole attention to our experience – to what we are seeing, feeling, tasting, smelling or hearing – rather than to our thoughts. It means living through our senses and our experience rather than through our thought-minds. For example, when you're having a shower in the morning, instead of letting your mind chatter away about the things you've got to do today or the things you did last night, bring your attention into the here and now. Be aware of the sensation of the water splashing against and running down your body and the sense of warmth and cleanness you feel. When you do chores such as mowing the lawn or washing the dishes, don't listen to music on your headphones or let yourself daydream. Give your attention to the objects and phenomena around you and to the physical sensations you are experiencing.

One thing you'll find is that such chores become more enjoyable. Mindfulness always brings a sense of wellbeing. This is partly because our perception becomes richer. Mindfulness *de-automatizes* our perceptions. Colours appear brighter, textures seem more intricate, scenes more beautiful and events more fascinating. While washing up, we're struck by the beauty of the reflections of light on the water and the bubbles that foam and pop around the plates and dishes. While mowing the lawn, we really notice the rich green colour of the grass and the perfect blueness of the sky above us.

This perceptual richness has a time-expanding effect. Mindfulness circumvents the psychological mechanism that switches off our attention to the reality of familiar experiences and environments. It therefore increases the amount of information that our minds process. As Jon Kabat-Zinn, one of the pioneers of the modern mindfulness movement, has written:

If you were really present with your moments as they were unfolding, no matter was what happening, you would discover that each moment is unique and novel and therefore momentous. Your experience of time would slow down. You might even find yourself stepping out of the subjective experience of time passing altogether, as you open to the timeless quality of the present moment . . . The slower the passage of time [becomes] from the point of view of your experience of it, the "longer" your life becomes, as you are here for more of your moments.[7]

So if we could make mindfulness our default state, or at least one that we frequently experience, it would have a similar time-expanding effect to travelling to unfamiliar environments.

Absorption and Time

There are also some straightforward implications of the third law of psychological time: "Time seems to speed up in states of absorption (or inversely, time seems to slow down in states of non-absorption – for example, boredom or impatience)."

However, perhaps we should start by looking at this law from the alternate perspective of speeding up rather than slowing time. There are many situations where we would prefer time to pass quickly – for example, on a long flight or train journey, in a waiting room or during a long shift at work. The key to speeding up time in these situations is absorption. On long journeys, we try to make long flights pass more quickly by reading, watching films, sleeping or working on a computer. A worker who wants her shift to pass quickly should cultivate a state of flow in which she loses self-awareness, along with awareness of time. Shop and bar workers often prefer their jobs to be busy, because busyness allows them to forget themselves and enter a state of flow, speeding up time.

Applying the same logic in reverse, if we don't want time to pass quickly then we should avoid states of absorption. If you have a day off work, don't spend it watching TV series or films – it will pass by in a flash. Instead, spend the day mindfully: walking and watching the world go by, talking to friends or gardening or exercising. If you want a week's vacation to last for as long as possible, don't do anything too absorbing. Don't spend your vacation sitting on the beach reading bestsellers or listening to podcasts. Don't go to a bar every evening to socialize or watch entertainment. Instead, expand time by walking through the countryside, exploring different towns, trying out new activities and chatting to people.

Of course, I'm not recommending that we avoid absorption entirely. Absorption can be very beneficial, especially in the form of flow. We need to regularly experience flow, especially in our working lives, to maintain a sense of wellbeing. But flow should be balanced with mindfulness. We should limit the amount of time we spend in absorption and ensure that we spend just as much time – if not more – in a mindful state. (Remember that, as mentioned in Chapter 1, flow and mindfulness are distinct. Mindfulness is a wide-ranging open awareness, whereas flow is a narrow state of immersion in one particular activity. This is why flow tends to make time pass quickly, while mindfulness tends to make it go slowly.)

Transcending Time

However, in this book we have been mainly focusing on time expansion of a different order of magnitude – not just slight variations caused by information processing, but radical changes that occur in different states of consciousness. In Chapter 1, this was represented by the fourth law of psychological time: "Time passes very slowly in intense altered states of consciousness, when our normal psychological structures and processes are significantly disrupted, our normal 'self-system' dissolves."

So is it possible to apply this fourth law in a similar way to the second and third laws? Can we induce time expansion experiences at will by cultivating altered states of consciousness? Perhaps we could even cultivate an *ongoing* altered state and slow down our *normal* perception of time, so that every day or week of our lives is expanded?

Of course, it would be incredibly useful to induce TEEs at will. Think about emergency workers. In the same way that people in accidents have more time to take preventative action, if paramedics or firefighters could slow down time in emergency situations, they would respond more quickly and effectively and so improve their chances of saving lives. In a similar way, surgeons could stretch time in operations. Less urgently, students could expand time in exams. Dentists and mechanics could stretch time whenever they are running behind schedule.

The ability to slow down time would be a huge benefit to athletes and sportspeople too. As we saw in Chapter 3, some athletes are able to do this – at least occasionally – which lifts their performance far above the level of their peers. Although they don't consciously try to slow down time, it follows naturally from a state of super-absorption. We also saw that a small number of athletes – such as the cricketer Don Bradman, the baseball player Ted Williams and the footballer Lionel Messi – may experience an *ongoing* state of time expansion, due to their soft self-boundaries or their susceptibility to altered states.

Obviously, consciously cultivating TEEs is a completely different proposition to slowing down time by enriching our perception or avoiding absorption. While time distortions due to the second and third laws are an everyday experience, TEEs are relatively rare, But the fact that some athletes have a natural facility for time expansion offers us some hope. After all, we know that this facility is due to factors such as super-absorption or a soft self-boundary. So perhaps we could cultivate these states through rigorous mental training.

What would this mental training be? The Japanese concept of *mushin* – literally, "no mind" – is helpful here. Mushin is a state of complete mental stillness in which our minds are free of emotions and discursive thoughts. It is a core concept of both Zen Buddhism and Japanese martial arts such as judo, karate and aikido. Martial artists cultivate mushin through intensive training, then apply it in combat and in their daily lives. (In Chinese Daoism, which resembles Zen in many ways, the equivalent concept is *wuxin*, which is cultivated through practices such as Qi Gong and Tai Chi.)

Many martial arts associate mushin with time dilation. As with the Zone state for athletes, it gives martial artists more time to perform intricate actions and to respond to an opponent's movements. In my research, a member of the Japanese martial arts school of Bujinkan described a TEE that occurred when he was taking a test to progress to a higher grade. In this test, the student kneels on the floor with their eyes closed, while the master stands above them with a bamboo sword, poised to strike. The student must sense the downstroke of the sword and roll out of the way at the right moment, not too early and before their head receives the blow. The man who took the test told me:

> I saw a silvery white bright line in my mind, like a column of light. I had a metallic sharp feeling. Then I heard a high-pitched tone, which felt like out-of-phase speakers on a stereo. I knew the sword was coming and I knew it was time to roll. I had lots of time to move out of the way and roll on to my side. It felt I had two or three seconds to think and position myself, from the time the sword came down toward me.

In reality, of course, the movement of the sword could only have lasted a split second. As the student told me, it is only possible to pass this test in a state of mushin, with an expanded sense of time. It is essentially a test of time expansion.

Similarly, in my earlier book *Making Time*, I referred to a sports coach called Mike Hall who believed that, after practising Tai Chi for 12 years, he was able to enter the Zone at will. He used the ability when playing squash, describing "a feeling of stillness, like I'm not trapped in sequential time anymore. The ball still darts around, but it moves around the court at different speeds depending on the circumstances."[8] Hall suggested that, in theory, it should be possible for any athlete to learn this ability through mental training.

Meditation

The best way to cultivate mushin – and therefore time expansion – is meditation. The main aim of all meditation practices is to attain a state of no-mind. All techniques of meditation, whether we focus on a mantra or our breathing or simply observe our own mental processes, aim to quiet our thought-chatter and cultivate mental emptiness. In a deep meditative state, the mind becomes so still and empty that our normal self-boundary becomes soft, bringing a sense of oneness with a pure consciousness that is both within us and beyond us. At this depth, there may also be a sense of time-transcendence. We may experience the "timeless moment" or "eternal now" that mystics like Meister Eckhart spoke of.

I have experienced this state many times, both during and after meditation. After a deep meditation, I find it difficult to estimate how long I have been practising, although it usually feels as if I have been "away" for a long time on a journey into a different level of reality. I feel a sense of ease, as my life is flowing gracefully in harmony with the world. When I look around my room, objects feel intensely vivid, more intricate and beautiful than normal. Sometimes I feel that I'm outside with them, as a part of the scene, rather than looking at them from a vantage point inside my own mind. There is also a sense of time expansion, if only relatively mild. I feel as if I'm moving more slowly through my life, with more time to do chores and perform tasks.

It's important to remember that meditation is also a long-term process. When we meditate regularly over a long period, we change our mental processes and structures on a permanent basis. Even during the periods when we aren't meditating, as we live through our daily lives, our minds are permanently quieter and emptier. Our self-boundary becomes permanently softer. This is why people often report that regular meditation enhances creativity and even increases susceptibility to psi experiences. And by the same logic, regular meditation should lead to a more expansive experience of time.

The idea that meditation can stretch time is supported by research. In 2013, a group of Romanian students practised half an hour of mindfulness meditation every day for a week. At the end of the week, they watched two short documentary films, each lasting for five minutes. They estimated the length of the films as significantly longer than a control group who hadn't practised meditation. On a long-term basis, in 2014 a group of German researchers interviewed 42 people who had been meditating regularly for many years. Compared to a similar sample of non-meditators, they felt less pressurized by time and reported a slower passage of time. As one of the researchers, Marc Wittmann summarized, the findings showed that experienced meditators "feel more time expansion . . . [L]ife as a whole passes more slowly and periods of time expand for people who live mindfully."[9]

Other Ways of Softening Our Self-Boundary

Regular meditation is, therefore, an important tool of time expansion. And this effect can be enhanced by living a generally meditative – or mindful – lifestyle. The great thing about mindfulness is that we can apply it to every aspect of our lives. We can eat breakfast mindfully, have a shower mindfully, walk to work mindfully, interact with other people mindfully, and so on. Mindfulness has a time-slowing effect in two different ways: in addition to softening our self-boundary, it increases information processing, by (as we saw above, see page 209) de-automatizing our perception.

Many simple everyday activities also have a meditative effect. In my book *The Adventure*, I use the term "active meditation" for physical activities like swimming, running, climbing and surfing. (This term could also be applied to psycho-physical exercises such as Yoga and Chi Gong.) Such activities can help to quieten our minds and to cultivate mushin. Walking outdoors is also very effective, especially among beautiful natural landscapes. The beauty of nature focuses our attention outside our minds and its stillness seeps into our being. After a while, our self-boundary softens and we feel a sense of connection to our surroundings, as if we share the same essence as the trees and hills around us and the sky and clouds above us.

More generally, it's important not to fill our lives with too much activity and stress. We should allow ourselves regular periods of quietness and solitude, where we turn off all electronic devices and simply sit or walk, observing and contemplating the world, shifting into a mode of *being*, rather than doing. It's very difficult to live in a meditative way when we feel oppressed with duties and demands and expend all our energy trying to deal with them. Excessive activity makes our minds over-active, which strengthens our self-boundary. The more active our minds are, the more separate we become from the world. But in quietness and solitude, our mental activity slows down. We disidentify with our superficial separate egos and reattune to deeper levels of our being, where we naturally connect and participate with the world around us.

Alongside this, if possible, I recommend making a conscious attempt to live *slowly*. Whereas rushing equates with stress and separation, slowness naturally generates mindfulness and connection. Of course, slowness isn't very compatible with modern life, with all its pressures and demands. But we all have a degree of control over how we respond to situations, particularly in our leisure time. We can choose not to rush. We can choose to slow down when we walk, cook or do our chores, and so on. There is a paradox here: since slowness helps to soften our self-boundary, it actually creates more time in our lives.

Let me recommend one final practice to soften our self-boundaries: altruism. When the main aim of our lives is to satisfy our own needs and desires, we become more disconnected from others and from the world in general. We become more identified with our own egos and more firmly rooted within our minds and bodies. In this way, selfishness strengthens our self-boundary. But when we help and serve others, we transcend self-centredness. We reach outside ourselves, into the mental space of other beings, forming an empathic connection, and so soften our self-boundary.

Any type of altruism has this effect, whether it's supporting friends and relatives, helping neighbours or less fortunate members of your community, even just complimenting colleagues and acquaintances. However, the most powerful type of altruism is to strangers, where there is no egoic motive and little possibility of them reciprocating. So I recommend that, as you go about your dialy life, put yourself on "altruism alert", ready to respond to any stranger who may need help – people who are hungry, distraught, lost, or who have falls or accidents.

Spirituality

All the practices and activities I've described above – and the overall lifestyle that would incorporate such activities – could be described as "spiritual". The essence of spirituality is to transcend separateness and cultivate connection. Spiritual development means becoming less ego-oriented and more connected to our own deeper being, to other beings, to nature and to the world in general. The aim of all spiritual traditions – including Hindu Yoga and Vedanta, Taoism, Sufism, the Kabbalah and Christian mystical traditions – is to undo our normal state of apparent separateness and realize our fundamental unity with the universe. To attain this goal, the traditions recommend a host of different practices and lifestyle guidelines. As well as ethical guidelines and specific psycho-spiritual techniques (such as breath control, *asanas*, sacred dancing or contemplation of sacred texts), they all recommend some form of meditation, as well as advising us to live quietly and simply, and to practise service and altruism.

In Chapter 5, where we investigated spiritual TEEs and CEEs, we briefly discussed cases of people becoming spiritually "awakened" after intense trauma and turmoil. We noted that one of the ongoing effects of their shift was an expansive sense of time. As one person reported, "When you're present all the time every day seems full. A day seems to last for such a long time."[10] One described an ongoing sense of timelessness, with an awareness that "Time cannot exist in my direct experience of life itself."[11] Another person described an awareness of the "eternal now" of the mystics: "an intense mind-boggling feeling of the past, the present and the future all existing at once, which again is very hard to explain".[12]

Along with the ongoing Zone states of some athletes, these examples confirm that it *is* possible to live in an ongoing state of time expansion, or even time cessation. While the above people underwent sudden awakenings following intense turmoil, it's also possible to undergo awakening gradually. In fact, this is exactly what will happen if you follow some of the practices I've described above. To cultivate an ongoing state of time expansion is precisely equivalent to a process of spiritual awakening. (For a more detailed and structured path of spiritual awakening, see my book *The Adventure: A Practical Guide to Spiritual Awakening*.)

An alternative route is to adopt one of the spiritual traditions I've just mentioned, such as Vedanta, Sufism or the Kabbalah, and follow the practices and guidelines that they incorporate. Of course, these traditions differ in some important ways. They have different views about the nature of human existence and reality. Some of them include concepts of God, while others (such as Buddhism) are not theistic at all. However, they all move in the same direction: away from separateness, toward connection and unity. The only issue with adopting such traditions is that, to a greater or lesser extent, they require us to adopt metaphysical beliefs – about reincarnation, God or different realms of existence, and so on – that we may not be comfortable with. This is why many modern spiritual seekers follow an eclectic, self-directed path, following certain practices without adhering to any tradition in particular.

Freedom from Time

I'm not claiming that these strategies will cultivate an ongoing state of intense time expansion of the same magnitude as emergency TEEs. It's important to be realistic. While I believe it is possible to cultivate intense TEEs on a *temporary* basis through rigorous mental training and deep meditation, it's probably more realistic to hope for an ongoing state of *mild* time expansion. After all, as we live our lives, we can't dispense with our self-boundary entirely. We need to have some sense of self, with some degree of boundary, to function in the world. If we didn't have a centre of gravity inside our minds and bodies, we wouldn't be able to meet the practical demands of our lives.

However, it stands to reason that if our normal sense of time is generated by the boundary of our self-system, then softening that boundary to some degree – even if only slightly – will have a time-slowing effect. And even a mild degree of time expansion would be massively beneficial to us.

Softening our self-boundary doesn't mean that we cease to exist as individuals. Even while our identity extends outside, connecting to the rest of the world, our centre of gravity can remain within our own mind and body. Although living *entirely* without a self-boundary would make it difficult to function in the world, living with a *softer* boundary actually makes life much easier. It removes the painful sense of separation and incompleteness that causes so much human suffering and generates so much pathological human behaviour. Instead, we feel a sense of belonging, of participation, of being at home in the world.

And in this state, we don't just transcend separateness but also our normal sense of fast-flowing linear time. In the same way that our identity is no longer trapped inside our mental space, we are no longer trapped in time. Time is no longer an enemy, running away from us, pressurising us from the future. We feel as at home in time as we do in the world. We feel *at one* with time in the same way that we feel at one with the world.

Of course, since you possess free will, it's up to you whether you follow the above practices and guidelines or not. But even if you don't engage with them, you can still be confident that, in times of emergency or crisis, time expansion experiences will arise spontaneously. You can still take comfort in the knowledge that, when you face life-threatening danger, time will expand sufficiently for you to take preventative action and so enhance your chances of survival. You can rest assured that you will respond with calm detachment and find the resilience and skill required to deal with the danger. You may even experience the great paradox that life-threatening danger can bring intense wellbeing. Even if you come close to death – even if you *do* die for a short time – you may well experience a state of acceptance and deep serenity. Should you be lucky enough to return from an encounter with death, the experience will permanently transform you and free you from fear of death.

Finally, in a more general sense, you can take comfort from the fact that linear time is an illusion. Although in everyday life you might feel oppressed by time – struggling to keep appointments, to meet deadlines, fighting a losing battle against the aging process – you can remind yourself that your oppressor is actually illusory. How can you be oppressed by an illusion? How can you be wholly enslaved when you know that freedom is possible? How can time defeat you, when at the deepest essence of your being you are timeless?

At the essence of reality, and at the essence of our beings, there is only oneness: oneness between us and the world, between us and each other, and between the past, present and future.

We are walking hurriedly along the beach in a straight line, with our eyes focused rigidly ahead, so that we pay no attention to the ocean beside us. Occasionally, in unusual situations and states of mind, we slow down and look to the side. We stop for a moment, then leave our path. We walk toward the waves and the ocean immerses us.

As we swim in the ocean, we feel one with it. We feel its enormity all around us, its panoramic presence stretching in all directions, incorporating all times as well as all places. We sense that reality is far too vast and complex to be limited to linear time. We sense that life in the ocean is much more authentic and more fulfilling than a narrow, blinkered existence on the straight path outside the ocean. We know that linear time only comes into existence when we step outside the ocean.

We can't live in the ocean, it's true. While we live in the world, we have to walk on solid ground. But as we walk, we can widen our vision and sense the panorama stretching around us. We can open ourselves to the ocean. While we walk, we can be one with the ocean. We can even walk in the ocean's shallow edge, both in and out of the water.

Even while we live in time, we can sense timelessness.

ACKNOWLEDGMENTS

I would like to thank everyone who provided reports of their time expansion/cessation experiences – the participants of my original research project and others who sent reports in response to my articles. Thanks also to my agent, Isabel Atherton, and the excellent publishing team at Watkins, who have been so supportive. I am also grateful to Paul Marshall, Marc Wittmann and Bernard Carr for some helpful comments and conversations. In a strange way, I feel as though I should also thank the lorry driver who caused the accident I described in the introduction. Although I don't feel grateful to him for endangering my life, this book probably wouldn't exist without the insight and inspiration I gained from the experience. As is often the case with psychology and philosophy, it's impossible to grasp the significance of a phenomenon without experiencing it first-hand.

REFERENCES

Chapter 1 – Why Time Seems to Pass at Different Speeds: A Preliminary Overview

1. James, 2023, Chapter XV.
2. *ibid.*
3. *ibid.*
4. *ibid.*
5. Ornstein, 1969.
6. Ornstein, 1969; Block & Read, 1978; Poynter, 1983.
7. Taylor, *Out of Time*, 2003.
8. Matthews & Beck, 2016.
9. Avni-Babad & Ritov, 2003.
10. Klein et al., 2003.
11. Wittmann, 2018, p.25.
12. *ibid.*, p.37.
13. Ogden, 2020.
14. Ogden et al., 2023.
15. Gibbon, 1977.
16. Zakay, 2012.
17. Buonomano, 2017, p.103.
18. *ibid.*, p.95.
19. *ibid.*, p.96.
20. Wittman, 2016, p.130.
21. Teghil et al., 2020.
22. Taylor, 2018.
23. Ogden et al., 2023.

24. In James, 2003, Chapter XV.
25. *ibid.*
26. Avni-Babad & Ritov, 2003.
27. *Tao Te Ching*, verse 28.
28. Csikszentmihalyi, 1992, p.58.
29. Ogden et al., 2023.

Chapter 2 – In the Danger Zone: Time Expansion in Accidents and Emergencies

1. In Noyes & Kletti, 1972, p.50.
2. *ibid.*, pp.46–7.
3. In Naim & Weibel, 2020, pp.37–8.
4. Scott, 2018, p.1.
5. Noyes & Kletti, 1976.
6. Hall, 1984.
7. Stetson et al., 2007.
8. Buonomano, 2017, p.72.

Chapter 3 – The Zone of Peak Performance: Time Expansion in Sport

1. In Brolin, 2017, p.23.
2. In Murphy & Whyte, 1995, p.107.
3. In Dossey, 1982, p.170.
4. In Murphy & Whyte, 1995, p. 42.
5. In Egenes, 2015.
6. In Brolin, 2017, p.23.
7. In Murphy & Whyte, 1995, p.108.
8. In Brolin, 2017, p.20.
9. In Brolin, 2010, p.7.
10. In Brolin, 2017, pp.141–2.
11. In Murphy & Whyte, 1995, p.108.
12. In Brolin, 2017, p.20.
13. Davis, 2000.
14. Dunstan & Heenan, 2014.

15. Bristow, 2018.
16. Thalbourne, 2009.
17. Parker, 2019.
18. Murphy & Whyte, 1995.

Chapter 4 – Time out of Mind: Spiritual Experiences

1. Taylor & Egeto-Szabo, 2017.
2. Taylor, 2011, p.4.
3. Taylor, 2011, p.4.
4. Marshall, 2015.
5. "Life and Doctrine of Saint Catherine of Genoa", Chapter XIV.
6. In Spencer, 1963, p.242.
7. In Happold, 1986, p.279.
8. *ibid.*
9. In Happold, 1986, p.278.
10. In Marshall, 2019, p.314.
11. Jeffries, 2024, pp.38–9.
12. Linares Gutiérrez at al., 2022.
13. Huxley, 1954, p.73.
14. *ibid.*
15. *ibid.*, p.27.
16. Ward, 1957, p.65.
17. *ibid.*, p 6.
18. Shanon, 2001, p.42.
19. *ibid.*
20. Peter Sjöstedt-Hughes, 2022, p.228.
21. Bayne & Carter, 2018, p.4.
22. Stephens Newell, 1972, p.381.
23. Ogden & Montgomery, 2012.
24. *ibid.*
25. *ibid.*
26. Taylor, 2021a, p.135.
27. Taylor, 2017, p.100.
28. Taylor, 2021a, p.142.

29. Unpublished account from my research archive.
30. *ibid.*
31. *ibid.*
32. *ibid.*

Chapter 5 – Deep Time: Time in Near-Death Experiences

1. Taylor, 2021a, p.119.
2. *ibid.*, p.121.
3. "Dr Mary Neal survived after being declared 'cold-to-the-touch dead'. This is what she saw."
4. *ibid.*
5. van Lommel, 2010.
6. Greyson, 2021.
7. Fenwick & Fenwick, 1996, p.72.
8. Greyson, 2021.
9. Alexander, 2015, p.17.
10. David Ditchfield, personal communication.
11. *ibid.*
12. In Bernstein, 2003, p.5.
13. Moorjani, 2014.
14. *ibid.*, p.63.
15. Long, 2014.
16. Bennett, 1997, p.60.
17. Greyson, 2012, pp.31–2
18. Ring, 2024.
19. *ibid.*
20. *ibid.*
21. Long, 2014.
22. Greyson, 2021, p.96.
23. Greyson, 2021.
24. Greyson & Holden, 2003.
25. Greyson, 2021.
26. Renz, p.40.
27. *ibid.*, p.15.

Chapter 6 – The Life Review

1. In Tymn, 2014.
2. Noyes & Kletti, 1972, p.50.
3. Fenwick & Fenwick, 1996.
4. Hicks, 2007, p.23.
5. van Lommel, 2024.
6. In Taylor, 2011, p.167.
7. In Greyson, 2021, p.40.
8. "Excerpts from PMH Atwater's 3 NDEs."
9. In Greyson, 2021, pp.40–41.
10. *ibid.*, p.42.
11. Katz et al., 2017, p.77.
12. In Greyson, 2021, p.42.
13. "NDE Accounts."
14. In Sartori & Walsh, 2017, p.32
15. In Taylor, 2021b.
16. van Lommel, 2024, p.24.
17. "Exceptional Experiences", https://www.nderf.org/Archives /exceptional.html
18. Blackmore, 1993.
19. Katz et al., 2017.
20. Mobbs & Watt, 2011.
21. Vincente et al., 2022.
22. Seger et al., 2023
23. Moorjani, 2014, p.63.

Chapter 7 – Toward the Timeless: When the Self Dissolves

1. Arstilla, 2012.
2. Craig, 2009.
3. Buonomano, 2017, p.70.
4. *ibid.*, p.72.
5. e.g., Thomson, "Woman of 24 Found to Have No Cerebellum in Her Brain." See also Taylor, 2018.

6. Nahm et al., 2012.
7. Moncrieff et al., 2022.
8. Pandya et al., 2012, p.640.
9. Hall, 1984, pp.135–6.
10. Piovesan at al., 2019, p.1157.
11. James, 1985, p.388.
12. Wittmann, 2018, p.27.
13. Hartocollis, 1983, p.17.
14. David Ditchfield, personal communication.
15. In Taylor, 2007, p.99.
16. Carroll et al., 2008, p.150.
17. Stanghellini et al., 2016, p.46.
18. El-Meligi, 1972, p.262.
19. Stanghellini et al., 2016, pp.50–51.
20. Drever, 2022, p.53.
21. *ibid.*
22. These issues relate to a wider discussion about the difference between "spiritual awakening" – or awakening experiences – and mental disorders, which I addressed in my earlier book *The Leap*. Particularly when it occurs suddenly and dramatically, spiritual awakening can cause some psychological disturbances. As a result, it is sometimes confused with psychosis or schizophrenia. Unfortunately, due to a lack of awareness of spiritual awakening within the psychiatric profession, some people may be diagnosed with mental disorders even though they are undergoing a positive transformational experience. In extreme cases, they may even be committed to institutions. However, the main difference between psychosis and awakening is that in the former, the normal self-system disintegrates and leaves a psychological vacuum, whereas in the latter the normal self-system is replaced by a new kind of self-system, with new psychological processes and functions.

To use another metaphor, you could compare it to a political revolution, in which a stable but oppressive government is overthrown. This often leads to a period of chaos and violence, in which the infrastructure of a country breaks down. However, in the ideal scenario, the revolution leads to a new era of harmonious democracy, with new fairer systems and structures that bring greater happiness.

Chapter 8 – The Future Is Written: Precognition and Retrocognition

1. Krohn, 2024, p.15.
2. *ibid.*, p.18.
3. *ibid.*, p.20.
4. *ibid.*, p.25.
5. *ibid.*, p.30.
6. Wahbeh et al., 2018.
7. Monterio de Barros, 2022.
8. Rhine, 1934/1997.
9. Fukada, 2002.
10. Adams, 2010, p.65.
11. Saltmarsh, 1932.
12. Knight, 2022, p.26.
13. *ibid.*, p.135.
14. *ibid.*, p.124.
15. *ibid.*, p.148.
16. *ibid.*, p.168.
17. *ibid.*, p.131.
18. James, 1896, p.884.
19. Savva & French, 2003.
20. "Swedenborg & Wesley", 2024.
21. See https://twitter.com/MikeyWelsh71/status/118305899 923259392?lang=en-GB
22. Ring, 2024.
23. See Wilson, 1978, pp.147–9.
24. Honorton & Ferrari, 1989.
25. Mossbridge at al., 2012.

26. Judd & Gawronski, 2011, p.406.
27. Bem at al., 2014.
28. Smith et al., 2014.
29. Müller at al., 2019.
30. Utts, 1996, p.118.
31. Utts, 2017, p.1379.
32. Roe, 2021, p.4.
33. Schäfer & Schwarz, 2019.
34. Rosnow & Rosenthal, 2003.
35. Osty, 1923, p.224.
36. Proust, *In Search of Lost Time*, p.51.
37. *ibid.*
38. Toynbee, *A Study in History*, pp.129–30.
39. *ibid.*, p.139.
40. Osty, 1923, p.31.
41. Saltmarsh, 1938, p.108.
42. Thouless, 1972, p.140.

Chapter 9 – Beyond Linear Time: Perspectives from Physics

1. Newton, 2024.
2. "Original letter from Isaac Newton to Richard Bentley, dated 17 January 1692/3", 2024.
3. Newton, 2024.
4. In Horgan, 1997, p.12.
5. In MacKenzie, 1997, p.124.
6. Carr, 2015a.
7. Callender, 2024.
8. Davies & Gribbin, 2007, p.14.
9. Liao et al., 2017.
10. In Mehra, 1973, p.244.
11. Rovelli, 2017, p.78.
12. Oreshkov et al., 2012.
13. Procopio et al., 2015.
14. Stapp, 2015.

15. Hawking & Mlodinow, 2010, p.140.
16. Cramer, 2016.
17. Rovelli, 2017, p.117.
18. Cardeña, 2018, p.663.
19. Reber & Alcock, 2018, p.391.
20. Carr, 2017.
21. Utts, 1995, p.1.
22. Carr, 2015b.
23. Utts, 1995, p.1
24. In Dossey, 1982, p.157.

Chapter 10 – Transcending Time: Softening the Self-Boundary

1. James, 1985, p.388.
2. Harris, 2012, p.2.
3. See Taylor, 2016.
4. In Rosenberg, 2021, p.91
5. Rhine, 1955, pp.20–21.
6. In Rosenberg, 2021, p.115.
7. Kabat-Zinn, 2005, p.162.
8. In Williams, 2006.
9. Wittmann, 2018, p.63.
10. Taylor, 2017, p.100.
11. Taylor, 2021a, p.142.
12. Unpublished account from my research archive.

BIBLIOGRAPHY

Adams, K, *Unseen Worlds*, Jessica Kingsley, London, 2010.

Alexander E, "Near-death experiences: The Mind-body debate & the nature of reality", *Mo Med.*, Jan–Feb, 112(1), 2015, pp.17–21.

Arstila, V, "Time slows down during accidents", *Frontiers in Psychology*, 3(196), 2012, doi: 10.3389/fpsyg.2012.00196

Article 196, 2012, www.frontiersin.org/article/10.3389/fpsyg.2012.00196

Avni-Babad, D & Ritov, I, "Routine and the Perception of Time", *Journal of Experimental Psychology*, 132(4), 2003, pp.543–50.

Bayne, T & Carter, O, "Dimensions of consciousness and the psychedelic state." *Neuroscience of Consciousness*, 2018(1), 2018, https://doi.org/10.1093/nc/niy008

Bem, D J, "Feeling the future: Experimental evidence for anomalous retroactive influences on cognition and affect", *Journal of Personality and Social Psychology*, 100, 2011, pp.407–25.

Bem, D, Tressoldi, P E, Rabeyron, T & Duggan, M, "Feeling the future: A meta-Analysis of 90 experiments on the anomalous anticipation of random future events", 2014, http://ssrn.com/abstract=2423692 or http://dx.doi.org/10.2139/ssrn.2423692

Bennett, R, *To Heaven and Back: True Stories of Those Who Have Made the Journey*, Grand Rapids, MI: Zondervan, 1997.

Block, R A & Reed, M A, "Remembered duration: Evidence for a contextual-change hypothesis", *Journal of Experimental Psychology: Human Learning and Memory*, 4(6), 1978, pp.656–65.

Bernstein, P, "Physicist uses NDEs to clarify the nature of time." *Vital Signs*, 22(2), 2003, pp.3–12, www.astro.sk/~msaniga/pub/ftp/intw-vs.pdf

Blackmore, S, *Dying to Live: Science and the Near-death Experience*, Grafton, London, 1993.

Buonomano, D, *Your Brain is a Time Machine: The Neuroscience and Physics of Time*, Norton, New York, 2017.

Bristow, T, "Rio Ferdinand attempts to explain Lionel Messi's magic after Barcelona star steals the show at Chelsea", *Daily Mirror*, www.mirror.co.uk/sport/football/news/rio-ferdinand-attempts-explain-lionel-12062009

Brolin, C, *Overdrive: Formula One in the Zone*, Vatersay Books, London, 2010.

Brolin, C, *In the Zone: How Champions Win and Think Big*, Blink Publishing, London, 2017.

Callender, C, "Is time an illusion?", *Scientific American*, 2010, www.scientificamerican.com/article/is-time-an-illusion/

Cardeña, E, "The experimental evidence for parapsychological phenomena: A review", *American Psychologist*, 73(5), 2018, pp.663–77, https://doi.org/10.1037/amp0000236

Carr, B J, "Hyperspatial models of matter and mind", in Kelly, E, Crabtree, A & Marshall, P (eds.), *Beyond physicalism: Toward reconciliation of science and spirituality*, 2015a, Rowman & Littlefield, London, pp.227–73.

Carr, B J, "Higher dimensions of space and time and their implications for psi", in May, E & Marwaha, S (eds.), *Extrasensory perception: Support, skepticim and science*, Vol. 2, 2015b, Greenwood Publishing, pp.21–61.

Carr, B, "Making Time and Space for Mind and Spirit", *Psychic News* (August), 2017, pp.38–40.

BIBLIOGRAPHY

Carroll, C et al., "Temporal processing dysfunction in schizophrenia", *Brain Cogn.*, 67(2), 2008, pp.150–61, doi: 10.1016/j.bandc.2007.12.005

Cooper, L & Erickson, M, *Time Distortion in Hypnosis*, Williams and Wilkins, Baltimore, 1959.

Craig, A D, "Emotional moments across time: a possible neural basis for time perception in the anterior insula", *Philos Trans R Soc Lond B Biol Sci.*, 64(1525), 2009, pp.1933–42, doi: 10.1098/rstb.2009.0008

Csikszentmihalyi, M, *Flow: The Psychology of Happiness*, Rider, London, 1992.

Davies, P & Gribbin, J, *The Matter Myth: Dramatic Discoveries that Challenge Our Understanding of Physical Reality*, Simon & Schuster, New York, 2007

Davis, C, *The Best of the Best*, ABC books, Sydney, 2000.

Dossey, L, *Space, Time and Medicine*, Shambhala, Boston, 1982.

"Dr Mary Neal survived after being declared 'cold-to-the-touch dead'. This is what she saw." www.mamamia.com.au/surviving-death-netflix/

Drever, M, *Reconceptualising dementia. Amelesia not dementia: Unmindfulness not madness, A manifesto*, Amelesia Books, London, 2022.

Dunstan, D & Heenan, T, "Shattering the Don Bradman Myth", *The New Daily*, 2022, www.thenewdaily.com.au/sport/2014/01/27/shattering-bradman-myth

Egenes, L, "Multitasking and the Transcendental Meditation Technique", 2015, https://tm-women.org/multitasking-and-the-transcendental-meditation-technique/

El-Melegi, A M, "Exploring Time in Mental Disorders", in Yaker, H, Osmond, H & Cheek, F (eds.), *The Future of Time*, Hogarth Press, London, 1972, pp.260–68.

"Exceptional Experiences", Near-Death Experience Research Foundation, www.nderf.org/Archives/exceptional.html

"Excerpts from PMH Atwater's 3 NDEs", https://ndestories.org/pmh-atwater

237

Fenwick, P & E, *The Truth in the Light*, Headline, London, 1996.

Fukuda, K, "Most experiences of precognitive dream could be regarded as a subtype of déjà-vu experiences", *Sleep and Hypnosis*, 4(3), 2002, pp.111–14.

Gibbon, J, "Scalar expectancy theory and Weber's law in animal timing", *Psychological Review*, 84 (3), 1977, pp.279–325, doi:10.1037/0033-295X.84.3.279

Greyson, B, *After*, St. Martin's Press, New York, 2021.

Hall, E, *The Dance of Life*, Anchor Press, New York, 1984.

Happold, F C, *Mysticism*, Pelican, London, 1986.

Harris, S, *Free Will*, Simon & Schuster, New York, 2012.

Hartocollis, P, *Time and Timelessness or the Varieties of Temporal Experience*, International Universities Press, New York, 1985.

Hawking, S & Mlodinow, L, *The Grand Design*, Bantam, London, 2010.

Hicks, G, *One Unknown*, Rodale, London, 2007.

Honorton, C & Ferrari, D C, "'Future telling': A meta-analysis of forced-choice precognition experiment, 1935–87", *Journal of Parapsychology*, 53, 1989, pp.281–308.

Horgan, J, *The End of Science: Facing the Limits of Knowledge in the Twilight of the Scientific Age*, Little, Brown, London, 1997.

Huxley, A, *The Doors of Perception* and *Heaven and Hell*, Penguin, London, 1988.

James, W, "Address of the president before the society for psychical research", *Science*, 3(77), 1896, pp.881–8.

James, W, *The Principles of Psychology*, 1890, www.yorku.ca/pclassic/James/Principles/

James, W, *The Varieties of Religious Experience*, Penguin, London, 1985.

Jeffries, R, *The Story of My Heart*, 1883, https://archive.org/details/storymyheartmya02jeffgoog/page/n3/mode/2up

Kabat-Zinn, J, *Coming to Our Senses*, Piatkus, London, 2005.

Katz, J et al., "The life review experience: Qualitative and quantitative characteristics", *Consciousness and*

Cognition, 48, 2017, pp.76–86, https://doi.org/10.1016/j.concog.2016.10.011

Klein L C, Corwin E J & Stine, M M, "Smoking Abstinence Impairs Time Estimation Accuracy", *Psychopharmacology Bulletin*, 37(1), 2003, pp.90–95.

Knight, S, *The Premonitions Bureau*, Faber, London, 2022.

Krohn, E, "The Eternal Life of Consciousness", 2021, www.bigelowinstitute.org/wp-content/uploads/2022/10/krohn-eternal-consciousness.pdf

Liao, S, et al., "Satellite-to-ground quantum key distribution", *Nature*, 549, 2017, pp.43–7, https://doi:10.1038/nature23655

"Life and Doctrine of Saint Catherine of Genoa", https://ccel.org/ccel/catherine_g/life/life

Linares Gutiérrez D et al., "Changes in subjective time and self during meditation", *Biology* 11(8), p.1116, doi: 10.3390/biology11081116.

Long, J, "Near-death experiences: Evidence for their reality", *Missouri Medicine*, 11(5), 2014, pp.372–80.

MacKenzie, A, *Adventures in Time*, Athlone Press, London, 1997.

Marshall, P, "Mystical experiences as windows on reality", in E F Kelly, A Crabtree & P Marshall (eds.), *Beyond Physicalism: Toward Reconciliation of Science and Spirituality*, Rowman & Littlefield, Lanham, MD, 2015, pp.39–78.

Marshall, P, *The Shape of the Soul*, Rowman and Littlefield, Lanham, MD, 2019.

Matthews, W J & Meck, W H, "Temporal cognition: Connecting subjective time to perception, attention, and memory", *Psychological Bulletin*, 142(8), 2016, pp.865–907, doi: 10.1037/bul0000045

Mehra, J, *The Physicist's Conception of Nature*, Reidel, Dordrecht, Holland, 1973.

Mobbs D & Watt C, "There is nothing paranormal about near-death experiences: how neuroscience can explain seeing bright lights, meeting the dead, or being convinced you are one of them", *Trends Cogn Sci.*, 15(10), 2011, pp.447–9, doi: 10.1016/j.tics.2011.07.010

Moncrieff, J et al., "The serotonin theory of depression: a systematic umbrella review of the evidence", *Mol Psychiatry* 28, 2023, pp.3243–56, https://doi.org/10.1038/s41380-022-01661-

Monteiro de Barros M C et al., "Prevalence of spiritual and religious experiences in the general population: A Brazilian nationwide study", *Transcultural Psychiatry*, April 2022, doi:10.1177/13634615221088701

Moorjani, A, *Dying to Be Me: My Journey from Cancer, to Near Death, to True Healing*, Hay House, Carlsbad, CA, 2014.

Mossbridge J, Tressoldi P & Utts J, "Predictive physiological anticipation preceding seemingly unpredictable stimuli: a meta-analysis", *Frontiers of Psychology*, 3(390), 2012, www.frontiersin.org/journals/psychology/articles/10.3389/fpsyg.2012.00390/full

Müller, M, Müller, L & Wittmann, M, "Predicting the Stock Market: An Associative Remote Viewing Study", *Zeitschrift fürAnomalistik Band 19*, 2019, pp.326–46.

Murphy, M & Whyte, R A, *In the Zone: Transcendent Experience in Sports*, Penguin, London, 1995.

Nahm M et al., "Terminal lucidity: a review and a case collection", *Arch Gerontol Geriatr.*, 55(1), 2011, pp.138–42. doi: 10.1016/j.archger.2011.06.031.

Nahm, M & Weibel, A, "The significance of autoscopies as a time marker for the occurrence of near-death experiences", *Journal of Near-Death Studies*, 38(1), 2020 doi: 10.17514/jnds-2020-38-1-p26-50

"NDE Accounts", https://iands.org/ndes/nde-stories/

Newell, S, "Chemical Modifiers of Time", in Yaker, H, Osmond, H & Cheek, F (eds.) *The Future of Time*, Hogarth Press, London, 1972, pp.378–81.

Newton, I, *The Mathematical Principles of Natural Philosophy*, https://en.wikisource.org/wiki/The_Mathematical_Principles_of_Natural_Philosophy, 2024.

Noyes, R & Kletti, R, "The experience of dying from falls", *OMEGA – Journal of Death and Dying*, 3(1), 1972, pp.45–52, https://doi.org/10.2190/96XL-RQE6-DDXR-DUD5

Noyes, R & Kletti, R, "Depersonalization in the face of life-threatening danger: A description", *Psychiatry*, 39(1), 1976, pp.19–27, https://doi.org/10.1080/00332747.1976.11023873

Ogden, R, "The passage of time during the UK Covid-19 lockdown", *PLoS One*, 15(7), 2020, https://journals.plos.org/plosone/article?id=10.1371/journal.pone.0235871

Ogden, R et al., "Distortions to the passage of time during chronic pain: A mixed method study", *European Journal of Pain*, 2023, https://doi.org/10.1002/ejp.2211

Ogden, R et al., "Changing experiences of the passage of time with age: do Christmas and Ramadan really come around more quickly each year?" Poster Presentation, European Congress of Psychology, 2023.

Ogden, R & Montgomery, C, "High Time", *The Psychologist*, 2012, www.bps.org.uk/psychologist/high-time

Oreshkov, O, Costa, F & Brukner, Č, "Quantum correlations with no causal order", *Nat Commun* 3(1092), 2012, https://doi.org/10.1038/ncomms2076

"Original letter from Isaac Newton to Richard Bentley", www.newtonproject.ox.ac.uk/view/texts/normalized/THEM00255, 2024.

Ornstein, R, *On the Experience of Time*, Penguin, London, 1969.

Osty, E, *Supernormal Faculties in Man: An Experimental Study*, E P Dutton & Co., New York, 1923.

Pandya, M et al., "Where in the brain is depression?" *Current Psychiatry Reports*, 14(6), pp.634–42, http://doi.org/10.1007/s11920-012-0322-7

Parker, S, "Training attention for conscious non-REM sleep: The yogic practice of yoga-nidrā and its implications for neuroscience research", *Progress in Brain Research*, 244, 2019, pp.255–72, https://doi.org/10.1016/bs.pbr.2018.10.016.

Piovesan, A et al., "The relationship between pain-induced autonomic arousal and perceived duration", *Emotion*, 19(7), 2019, pp.1148–61, doi:10.1037/emo0000512

Poynter, W D, "Duration judgment and the segmentation of experience", *Memory & Cognition*, 11(1), 1983, pp.77–82.

Procopio, L et al.,"Experimental superposition of orders of quantum gates", *Nat Commun* 6(7913), 2015, https://doi.org/10.1038/ncomms8913

Proust, M, *In Search of Lost Time, Volume 1* (trans. C K Scott Moncrieff & T Kilmartin), Chatto & Windus, London, 1992.

Reber, A S & Alcock, J E, "Searching for the impossible: Parapsychology's elusive quest", *American Psychologist*, 75(3), pp.391–9, https://doi.org/10.1037/amp0000486

Renz, M, *Dying: A Transition.* Columbia University Press, New York, 2015.

Rhine, J B, *Extra-Sensory Perception*, Branden, Boston, MA, 1934/1997.

Rhine, L E, "Precognition and intervention", *Journal of Parapsychology*, 19, 1955, pp.1–34.

Ring, K, "The Life Review in Reverse: Visions of the Future", 2022, www.lifeafterlife.com/blog/the-life-review-in-reverse-visions-of-the-future

Roe, C, "Small Wonder: Effect Sizes in Parapsychology", *Magazine of the Society of Psychical Research*, 2021, pp.4–5.

Rosenberg, B, "Precognition", in Kelly, E & Marshall, P. (eds.), *Consciousness unbound: Liberating mind from the tyranny of materialism*, 2021, Rowan & Littlefield, London, pp.89–138.

Rosnow, R L & Rosenthal, R, "Effect sizes for experimenting psychologists", *Canadian Journal of Experimental Psychology/Revue canadienne de psychologie expérimentale*, 57(3), 2003, pp.221–37, https://doi.org/10.1037/h0087427

Rovelli, C, *The Order of Time*, Penguin, London, 2017.

Saltmarsh, H F, "Report of Cases of Apparent Precognition", *Proc. SPR*, 42, 1934, pp.49–103.

Saltmarsh, H F, *Foreknowledge*, G Bell & Sons, London, 1938.

Sartori, P & Walsh, K, *The Transformative Power of Near-Death Experiences: How the Messages of NDEs Positively Impact the World*, Watkins, London, 2017.

Savva, L & French, C, "An investigation into precognitive dreaming: David Mandell – the man who paints the future?" *Paper presented at the Annual Conference of the Society for Psychical Research*, Manchester Metropolitan University, 2003.

Schäfer, T & Schwarz, M A, "The meaningfulness of effect sizes in psychological research: Differences between sub-disciplines and the impact of potential biases", *Frontiers in psychology*, 10(813), 2019, https://doi.org/10.3389/fpsyg.2019.00813

Scott, D, *Up and About: The Hard Road to Everest*, Veterbrate Publishing, Sheffield, 2018.

Seger, S E, Kriegel, J L, Lega, B C & Ekstrom, A D "Memory-related processing is the primary driver of human hippocampal theta oscillations", *Neuron*, 2023, doi: 10.1016/j.neuron.2023.06.015

Shanon, "Altered temporality", *Journal of Consciousness Studies*, 8(1), 2001, pp.35–58.

Sjöstedt-Hughes, P, "The white sun of substance: spinozism and the psychedelic amor dei intellectualis", in Hauskeller, C & Sjöstedt-Hughes, P, (eds.), *Philosophy and Psychedelics: Frameworks for Exceptional Experience*, Bloomsbury Academic, London, 2022, pp.211–35.

Smith, C C, Laham, D & Moddel, G, "Stock market prediction using associative remote viewing by inexperienced remote viewers" *Journal of Scientific Exploration*, 28(1), 2014, https://journalofscientificexploration.org/index.php/jse/article/view/608

Spencer, S, *Mysticism*, London, Penguin, 1963.

Stapp, H, "A quantum-mechanical theory of the mind/brain connection", in Kelly, E F, Crabtree, A & Marshall, P (eds.), *Beyond Physicalism: Toward a Reconciliation of Science and Spirituality*, Rowman & Littlefield, Lanham, MD, 2015, pp.157–94.

Stetson, C, Fiesta, M & Eagleman, D, "Does time really slow down during a frightening event?" PLOS ONE, 2(12), 2007, https://doi.org/10.1371/journal.pone.0001295

Stanghellini, G et al., "Psychopathology of Lived Time: Abnormal Time Experience in Persons With Schizophrenia", *Schizophrenia Bulletin*, 42(1), 2016, pp.45–55, https://doi.org/10.1093/schbul/sbv052

"Swedenborg & Wesley", www.swedenborgstudy.com/articles/Swedenborgs-revelation/od058.htm

Taylor, S, *Out of Time*, Pauper's Press, Nottingham, 2003.

Taylor, S, *Making Time: Why Time Seems to Pass at Different Speeds and How to Control it*, Icon, London, 2007.

Taylor, S, *Out of the Darkness*, London, Hay House, 2011.

Taylor, S, "Reclaiming Human Freedom", *Philosophy Now*, 112, 2016, pp.15–17.

Taylor, S, *Spiritual Science: Why Science Needs Spirituality to Make Sense of the World*, Watkins, London, 2018.

Taylor, S, *Extraordinary Awakenings*, New World Library, Carlsbad, CA, 2021a.

Taylor, S, "'My life flashed before my eyes': a psychologist's take on what might be happening", *The Conversation*, 2021b, https://theconversation.com/my-life-flashed-before-my-eyes-a-psychologists-take-on-what-might-be-happening-162320

Taylor, S & Egeto-Szabo, K, 2017, "Exploring Awakening Experiences: A Study of Awakening Experiences in terms of their Triggers, Characteristics, Duration and After-Effects", *Journal of Transpersonal Psychology*, 49(1), 2017, pp.45–65.

Teghil, A et al., "Duration reproduction in regular and irregular contexts after unilateral brain damage: Evidence from voxel-based lesion-symptom mapping and atlas-based hodological analysis", *Neuropsychologia*,

147, 2020, www.sciencedirect.com/science/article/abs/pii/S0028393220302505

Thalbourne, M A, "Transliminality, anomalous belief and experience, and hypnotizability", *Australian Journal of Clinical and Experimental Hypnosis*, 37, 2009, pp.45–56.

Thomson, H, "Woman of 24 Found to Have No Cerebellum in Her Brain", *New Scientist*, 10 September 2014, 2014, www.newscientist.com/article/mg22329861-900-woman-of-24-found-to-have-no-cerebellum-in-her-brain

Thouless, R, *From Anecdote to Experiment in Psychical Research*, Routledge & Kegan Paul, London,1972.

Toynbee, A, *A Study in History, Vol. X*, Oxford University Press, London, 1954.

Tymn, M, "Admiral tells of drowning and what happened after", 2010, https://whitecrowbooks.com/michaeltymn/entry/admiral_tells_of_drowning_and_what_happened_after

Utts, J M, "An assessment of the evidence for psychic functioning", *Journal of Scientific Exploration*, 10(1), 1996, pp.3–30.

Utts, J M, "Appreciating Statistics", *Journal of the American Statistical Association*, 111, 2017, pp.1373–80, www.tandfonline.com/doi/full/10.1080/01621459.2016.1250592.

Utts, J M, "An Assessment of the Evidence for Psychic Functioning", Stargate Collection, CREST, CIA-RDP96-00791R000200070001-9, 1995 (Approved for Release 2000), www.cia.gov/readingroom/document/cia-rdp96-00791r000200070001-9

Van Lommel, P, *Consciousness Beyond Life: The Science of the Near-death Experience*, HarperOne, New York, 2010.

Van Lommel, P, "The continuity of consciousness: A concept based on scientific research on near-death experiences during cardiac arrest", 2021, www.bigelowinstitute.org/wp-content/uploads/2022/10/lommel-continuity-consciousness.pdf.

Vicente, R et al., "Enhanced interplay of neuronal coherence and coupling in the dying human brain", *Front. Aging Neurosci.*, 14, doi:10.3389/fnagi.2022.813531

Wahbeh, H, Radin, D, Mossbridge, J, Vieten C & Delorme A, "Exceptional experiences reported by scientists and engineers", *Explore*, 14(5), 2018, pp.329–41, doi: 10.1016/j.explore.2018.05.002. Epub 2 August 2018, PMID: 30415782.

Ward, R, *A Drug-Taker's Notes*, Victor Gollancz, London, 1957.

Wargo, E, *Time Loops*, Anomalist Books, San Antonio, TX, 2018.

Williams, C, "The 25 Hour Day", *New Scientist*, 4 February 2006.

Wilson, C, *Mysteries*, Granada, London, 1978.

Wittmann, M, *Altered States of Consciousness*, MIT Press, Cambridge, MA, 2018.

Wittmann, M, *Felt Time*, MIT Press, Cambridge, MA, 2017.

Zakay, D, "Experiencing time in daily life", *The Psychologist*, 25, 2012, pp.578–81.